T0122192

Projected Capacitive Touch

Tony Gray

Projected Capacitive Touch

A Practical Guide for Engineers

Tony Gray
Plano, TX, USA

ISBN 978-3-030-07490-6   ISBN 978-3-319-98392-9   (eBook)
https://doi.org/10.1007/978-3-319-98392-9

This Springer imprint is published by the registered company Springer Nature Switzerland AG
The registered company address is: Gewerbestrasse 11, 6330 Cham, Switzerland

Contents

About the Author

Tony Gray Tony has been designing, tuning, and testing projected capacitive touch panels since 2009. In his role as Director of PCAP Touch Technology first at Ocular LCD and then at Dawar Technologies, he has been responsible for over 100 custom projected capacitive touch panel design projects. Tony is highly regarded in the marketplace for his knowledge of touch systems. He has presented at several different technical conferences on projected capacitive touch technology and published numerous articles and white papers. Before working in the touch industry, Tony spent 15 years writing embedded software for a variety of devices including residential thermostats, smoke opacity meters, hospital food delivery systems, robotic assay equipment, and a system used to capture and analyze 3D neuronal structures. Tony has a Bachelor of Science in Computer Engineering from Lehigh University. He lives in Plano, Texas.

Chapter 1
Introduction

On June 29, 2007, the world changed forever. That was the day the first-generation iPhone® was released to the public. There are several reasons that the iPhone had such a huge impact. It redefined what a cell phone could do. It looked sleek, modern, and (mostly) seamless. It had a high-resolution full color display. It included a fully functional music player. It had an eco-system of custom applications. But the most important thing about the iPhone was that it could recognize two separate touches simultaneously. That may not sound like a major accomplishment today, but in 2007 it was revolutionary.

Is that picture of Grandma and the kids too small? Just use two fingers to zoom in and see the detail. Want to see how close Burkburnett Texas is to Dallas? Just use two fingers to zoom out until you can see both. Is that map of the zoo not oriented to the direction you are walking? Just use two fingers to rotate it so that it lines up.

Before the iPhone, the touch market was dominated by resistive touch sensors. Resistive touch sensors were cheap, but they also scratched easily, had to be replaced often, had to be calibrated on a regular basis, required a clunky bezel design, and only supported one touch (there were some multi-touch resistive solutions on the market, but they never worked well enough to capture significant market share). Several other touch technologies also had significant market share including surface acoustic wave and infrared, but resistive was the market leader for most applications. But once the iPhone was released, it established a new expectation for the touch experience.

I do not mean to imply that Apple® invented multi-touch. There were several multi-touch technologies available going back to the 1980s. Apple did not even invent projected capacitive multi-touch. That had been around since at least 1984 when it was used to create a 16 key capacitive touch pad, although it was not called "projected capacitive" until 1999.

Apple's true innovation was not the technology, but the use of it. The iPhone gave you a reason to need multiple touches. Once users had this capability and understood how useful and intuitive it was, they began to expect it on other devices as well from

© Springer International Publishing AG, part of Springer Nature 2019
T. Gray, *Projected Capacitive Touch*, https://doi.org/10.1007/978-3-319-98392-9_1

thermostats to exercise bikes to infusion pumps to slot machines to pretty much everything.

While the iPhone deservedly gets a lot of credit for introducing projected capacitive (PCAP) touch technology to the world, there were several other factors that also contributed to the rise of PCAP. Full color displays started getting less expensive, allowing designers to replace monochrome, segmented displays with full color bitmap displays. Processors continued to get less expensive, which was important because it is just about impossible to drive a 320×480 256 color TFT with an 8051 embedded processor and no operating system. Speaking of operating systems, another important factor was the continued growth of Linux in the embedded systems market. All of these changes fed into one another allowing product designers to put together inexpensive full color displays, a free operating system, and a high-end processor capable of displaying a multicolor GUI.

When all this upheaval started happening in 2007, I was working for Invensys Controls writing software for residential thermostats using a segmented LCD, a 16-bit processor, and no operating system. In 2007, our design team finally convinced marketing that we should develop a high-end thermostat with a 5.7in TFT and a resistive touch sensor. This was my first glimpse of the changes coming in product design.

In 2008, I left Invensys Controls to work for Ocular LCD. Ocular had been very successful supplying custom segmented LCDs to various embedded markets. But fairly soon after I joined the company, it became obvious that segmented LCDs were going to be replaced by full color TFTs in just a few years. My boss Larry Mozdzyn, the CTO and one of the co-founders of the company, realized that the same equipment we used to build LCDs could also be used to make projected capacitive touch sensors. Over the next few years Ocular transformed itself from an LCD supplier to a major supplier of PCAP touch sensors. And fortunately for me, Larry pulled me into the PCAP side of the business.

At the time I did not know anything about PCAP, so I did what any nerd would do: I Googled "how does projected capacitive touch work." I found a few web sites that did not go into a lot of detail. Then I tried searching Amazon® for a book on PCAP. I was a bit surprised that I could not find one, although this was not too shocking since PCAP was just starting to become popular. I attended trainings offered by the touch controller vendors we used at the time, Cirque® and Atmel®. I asked *a lot* of questions (no doubt to the point of annoying the engineers leading the trainings). I read every bit of documentation I could get. I printed out the documents, wrote notes on each page, and then e-mailed dozens of questions to the support teams at Cirque and Atmel. I wrote a lot of utilities for the controllers including configuration apps, test apps, programming apps, and even tuning wizards that automated the tuning process. Eventually I developed an understanding of PCAP including the fundamental physics of the technology, the operating principles of a touch controller, communication protocols and driver support, accuracy and linearity testing, physical construction, and manufacturing.

Occasionally, I would do another search on Amazon looking for the definitive book on projected capacitive touch technology, but I never found one. I had written a few articles over the years, some on embedded software development and some on PCAP. Eventually, it occurred to me that if nobody else was going to write a book on PCAP, I should do it.

This book is essentially the book I wish I had found on Amazon in 2008. It explains just about everything there is to know about projected capacitive touch that is not proprietary (if you are looking for a book to explain the secrets of the *insert-name-here* touch controller, sorry but NDAs prevent me from getting into that kind of detail). This book starts with a basic explanation of how a projected capacitive touch sensor works. It explains mutual vs. self capacitance, how capacitance is measured, how a touch sensor can be modeled as an electronic circuit, optical quality metrics, transparent conductors, and so much more (check the Table of Contents for a complete list of topics). In short, this book includes just about everything I have learned about PCAP over the last 10 years that I am allowed to share. While it is targeted at mechanical, electrical, and software engineers who are considering using PCAP in a project, I am hopeful that it will also prove useful to a wide variety of people from sales and marketing to PCAP designers to physics students to nerds who just want to learn more (my people). I hope you all enjoy reading it as much as I did writing it.

Chapter 2
Projected Capacitive Touch Basics

A capacitor is an energy storage device, a very simple form of battery. We form a capacitor by placing two conductive plates very close to each other (Fig. 2.1).

One plate of the capacitor is typically connected to ground and the other plate is connected to a voltage source through an in-line resistor. Over time the capacitor charges up to the applied voltage according to the equation:

$$V_{CAP} = V_{IN} \left(1 - e^{\frac{-t}{RC}} \right) \qquad (2.1)$$

t is time in seconds, R is the value of the resistor in Ohms, and C is the value of the capacitor in Farads. The voltage on the capacitor, V_{CAP}, charges up at an exponential rate (Fig. 2.2).

Notice that the axes on Fig. 2.2 do not have a scale. The general charging curve for a capacitor always looks like the graph shown. The Y scale is based on the charging voltage. The capacitor voltage approaches the applied voltage asymptotically (i.e., it will take an infinite time to reach the applied voltage). In theory, a capacitor can be charged to any voltage. In practice, there are physical constraints on the maximum voltage a particular capacitor can be charged to before breaking down, exploding, or failing in some other catastrophic and exciting way.

The scale of the X or time axis depends on the value of the capacitor which depends on the physical characteristics of the capacitor (it also depends on the resistor, but for now we are considering it to be a constant). In an ideal capacitor the important characteristics are the area of the plates, the distance between the plates, and the dielectric of the material between the plates. Think of the dielectric as a measurement of how well an electric field propagates through the material. Some materials do a better job of propagating electric fields and thus have a higher dielectric. Some materials do a worse job and have a lower dielectric.

The capacitance can be calculated from the area, distance, and dielectric (ε) using Eq. 2.2.

T. Gray, *Projected Capacitive Touch*, https://doi.org/10.1007/978-3-319-98392-9_2

Fig. 2.1 Capacitor

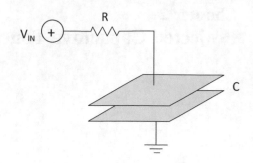

Fig. 2.2 Capacitor charging

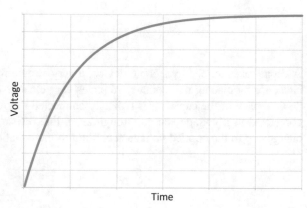

$$C = \varepsilon \frac{A}{d} \qquad (2.2)$$

The units for capacitance are Farads. One Farad is defined as the capacitance needed to store one Coulomb of charge with one volt across the plates.

A smaller capacitor charges very quickly; a larger capacitor charges very slowly. Most capacitors used in electronics have very small values in the micro Farad (10^{-6}) to nano Farad (10^{-9}) range. In projected capacitive touch sensors, we have to measure capacitances in the pico Farad range (10^{-12}).

A useful analogy is to think of the capacitor as a coffee mug and the current from the voltage source as coffee being poured into the mug (Fig. 2.3).

The applied voltage (V_{IN}) is the height of the mug, and the current height of the coffee in the mug is the current voltage on the positive plate of the capacitor (V_{CAP}). As you pour the coffee into the cup, you naturally begin to slow the rate of pouring as the cup gets more full. The capacitor is fully charged when the cup is completely filled.

If the mug is small, it gets filled very quickly. If the mug is large, it gets filled more slowly. The resistor in the circuit controls how fast the coffee comes out of the pot. Once the mug is full, you cannot fill it up any further and the flow of coffee stops. To extend the analogy a bit further, there is always some small leakage current from a capacitor. Think of this as a tiny hole in the bottom of the mug that drips

Fig. 2.3 Charge time analogy

coffee out very slowly. If you stop pouring coffee in the mug and wait long enough, the mug will eventually empty by itself. The same thing happens to capacitors. Once you charge a capacitor, it will slowly discharge over time.

As we pour coffee into the mug, we can measure the pour rate (current), the height of the coffee in the mug (voltage on the positive plate), and how long it takes to fill the mug (charging time). And of course we already know the maximum amount of coffee we can fit (applied voltage).

We said before that the capacitor never quite gets to the applied voltage level (i.e., the mug never completely fills up). So how do we determine when it is full enough? We need some standard way to define how full our capacitor is. To simplify matters, it would be great if the thing we use to measure how full it is could be ratiometric, like 50% or 70% of the full charge. That will allow us to have a standard metric for our discussions that we can all agree on, but that does not depend on the value of the capacitor or the voltage we are using.

The metric we use is called the time constant and is identified by the Greek letter Tau (τ). The time constant is defined as:

$$\tau = RC \tag{2.3}$$

The units for the time constant, τ, is seconds. That implies that multiplying resistance times capacitance gives time, which can be a bit surprising if you have never seen this equation before. We can see how this works out if we look at resistance (Ohms) and capacitance (Farads) in fundamental units:

$$\text{Ohms} = \frac{\text{kg} * \text{m}^2}{s^3 * A^2} \tag{2.4}$$

Table 2.1 Time constant levels

Tau	Equation	Charge level
1	$1 - e^{-1}$	63%
2	$1 - e^{-2}$	86%
3	$1 - e^{-3}$	95%
4	$1 - e^{-4}$	98%
5	$1 - e^{-5}$	99%

Fig. 2.4 Charge levels for τ

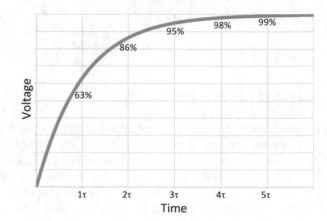

$$\text{Farads} = \frac{s^4 * A^2}{\text{kg} * \text{m}^2} \tag{2.5}$$

$$\text{Ohms} * \text{Farads} = \frac{\text{kg} * \text{m}^2}{s^3 * A^2} * \frac{s^4 * A^2}{\text{kg} * \text{m}^2} = s \tag{2.6}$$

Looking back at the equation for charging a capacitor:

$$V_{\text{CAP}} = V_{\text{IN}} \left(1 - e^{\frac{-t}{RC}}\right) \tag{2.7}$$

It is clear that the shape of the charge graph is defined by the values R and C. For that reason, τ is often called the RC time constant and the charging equation is often written like Eq. 2.8:

$$V_{\text{CAP}} = V_{\text{IN}} \left(1 - e^{\frac{-t}{\tau}}\right) \tag{2.8}$$

The standard way of defining the charge time is to use a multiplier of τ, as in 2τ, 3τ, 4τ, etc. Since τ is RC, one τ is $1*RC$, two τ is $2*RC$, and so on. So 2τ means when $t = 2 * RC$, or when two times RC seconds has passed. Table 2.1 shows the correspondence between different multipliers of τ and the actual charge level as a percentage of the applied voltage.

Figure 2.4 shows the charge graph for charge levels 1τ to 5τ.

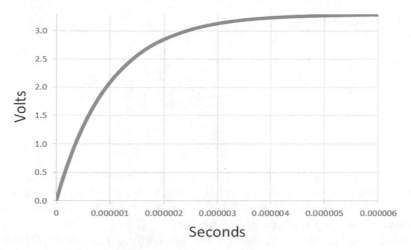

Fig. 2.5 Charge time for 10 pF and 100 kΩ

Let's start using a real example and come up with some real numbers. We will learn later that the capacitance values we are interested in are on the order of 1–10 pF (pico is 10^{-12}), so we will use a capacitor of 10 pF with a resistance of 100 kΩ and a charging voltage of 3.3 V. For this example, the charge graph looks like Fig. 2.5.

A capacitor is generally considered fully charged at 5τ or 0.9913 * 3.3 V = 3.27129 V. Our 5τ time for this circuit is 5 * 10 pF * 100 kΩ = 0.000005 s or 5 μs.

Now imagine that we do not know the capacitance and we want to determine it. This is an important point because the whole idea behind projected capacitive touch sensors is that we need to measure an unknown capacitance. The easiest way to do that is to use the charge time of the capacitor to determine its value. There are a few practical ways we can do this. One way would be to apply the 3.3 V, measure the amount of time it takes to charge up to 5τ, reverse Eq. 2.8, and use it to calculate the capacitance:

$$C = \frac{-t}{R * \ln\left(1 - \frac{V_C}{V_{IN}}\right)} \tag{2.9}$$

As an example, let's use the same numbers we found above for when our 10 pF cap is charged to 5τ:

$$C = \frac{-0.000005}{100000 * \ln\left(1 - \frac{3.27129}{3.3}\right)} = 10 * 10^{-12} = 10 \text{ pF} \tag{2.10}$$

While the math for this method is simple, the implementation is difficult. With the various setups, multiple sampling, averaging, etc. that we need to do, this measurement might take 1 μs or it might take 100 μs depending on the capacitance. It is

Fig. 2.6 Capacitor with
grounded rod

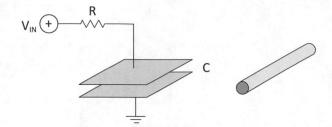

difficult to write reliable embedded software when the time for a measurement cycle
is unknown. And since we are eventually going to use this to locate fingers on a
touch sensor, we do not want the time between our touch reports to vary based on the
distance of the touch from one of the capacitors.

Another option is to charge the capacitor for a set time, then measure the voltage
on the charging plate. This has the benefit of always taking the same amount of time.
And we can solve it using the same equation as before. For example, if we measure
the voltage on the capacitor 1.85 μs after connecting the voltage source to the plate,
we measure 2.78 V. Our calculation becomes

$$C = \frac{-0.00000185}{100000 * \ln\left(1 - \frac{2.78}{3.3}\right)} = 10*10^{-12} = 10 \text{ pF} \qquad (2.11)$$

While measuring the voltage at the positive plate works from a theoretical point, it
does not really work with the physical construction of a touch sensor. Once we have
discussed how a touch sensor is actually built, we will come back to the issue of how
we actually determine a capacitor's value.

Now that we have a basic understanding of how to measure the value of a
capacitor, let's put it to work to do something useful. We will start with a question:
What happens if we place a grounded copper rod next to the capacitor? (Fig. 2.6).

When we begin dumping charge into the capacitor, some of that charge is going
to couple to the grounded rod and disappear into ground. Going back to the coffee
mug analogy, the grounded rod essentially acts like a second coffee mug. Most of the
coffee goes into the first mug, but some is diverted into the second mug (Fig. 2.7).

Because some of the charge is diverted, it takes longer to charge up the capacitor.
And since it takes longer to charge up the capacitor, the charge graph now looks like
Fig. 2.8.

If we use the same measurement technique to determine the capacitance, we find
that the value of the capacitor has increased (i.e., the coffee mug appears bigger than
it actually is). What is really going on is that we have added two new capacitors. The
first is from the positive capacitor plate to ground, and the second is from the bottom
capacitor plate to ground. This will be explained further when we get into the
equivalent circuits involved in a touch sensor. For now, just think of it as the
capacitor value increasing.

Fig. 2.7 Capacitor with grounded rod analogy

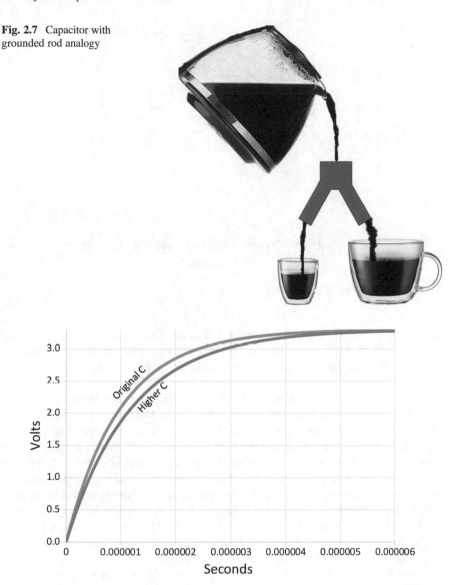

Fig. 2.8 Charge time with grounded rod

When the rod is closer to the capacitor, more charge is lost to ground, it takes longer to charge up the capacitor, and the measured capacitor value increases. When the rod is further away, less charge is lost and the measured capacitor value decreases. In theory, we can use this change in capacitance to calculate the distance between the metal rod and the capacitor (Fig. 2.9).

We just need to know the value of the capacitor without the metal rod in place, then the value of the capacitor with the metal rod in place. In practice, this is actually

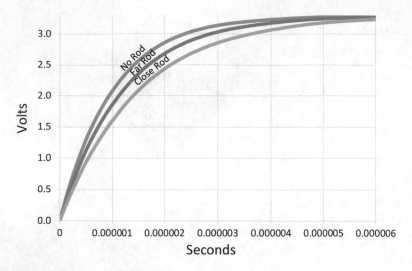

Fig. 2.9 Charge time varies by distance

quite difficult due to the complex interaction of the electric fields involved. But the basic idea, that we can use the change in capacitance to determine the relative distance between the capacitor and the rod, is true regardless of the messiness of the math.

When we say the metal rod is grounded, what does that mean? When something is truly grounded, it is literally inserted into the Earth. The reason for this is that, in theory, the Earth can supply a nearly infinite number of electrons, or act as a sink for a nearly infinite number of electrons (which can also be thought of as a nearly infinite number of electron holes). In reality, of course, the number of electrons or holes is not infinite, but it is much, much larger than the typical electrical circuit needs. In essence, by saying something is grounded we are saying that the ground can supply all the electrons or all the holes needed for the circuit to do whatever it is going to do.

When a printed circuit board is built, it typically has a ground plane which is a large copper area. This copper area is usually not connected to Earth ground (it may be connected to a grounded outlet which is connected to a power framework in the building which is connected to a terminal box which is eventually connected to the Earth, but from a practical stand point the copper is so far from Earth that it does not really matter). Even though the copper plane is not connected directly to Earth, it has way more electrons and holes than the circuit could ever possibly need assuming the circuit is operating normally. To the circuit, the copper plane looks like ground. This is called a virtual ground because it can supply all of the electrons or holes needed even though it is not actually Earth ground.

Now for the interesting part: a human body can also act as a virtual ground. For small enough circuits, the human body can supply enough electrons or holes that it appears to be ground. And since the body is a virtual ground, it can also couple to a circuit through an electric field. In other words, if we take away the metal rod and put

Fig. 2.10 Grid of capacitors

Fig. 2.11 Grid of capacitors with finger

our finger near the capacitor described above, some of the applied voltage will couple to our finger which changes the measured capacitance. We can use that change in capacitance to calculate the distance from the capacitor to our finger.

Next, imagine that we create a grid of capacitors as shown in Fig. 2.10 (the resistors and wires are omitted for clarity).

Each capacitor will have a slightly different value due to physical differences, proximity to the other capacitors, etc. The capacitors on the edges will have very different values since they are not surrounded by other capacitors. This will be important later when we talk about edge effects on touch sensors.

We measure the capacitance of each capacitor in the grid. Then we place a finger somewhere near the grid. Capacitors that are very close to our finger will have large changes in values (red in Fig. 2.11). Capacitors slightly further away will change less (blue). Capacitors still further from the finger will not be affected at all, or at least not much compared to those near the finger (gray). By measuring the changes in capacitances, and using a little interpolation, we can determine which capacitors the finger is closest to.

Our next task is to turn that into a useful coordinate. Let's assume that our rows and columns of capacitors are 6 mm apart (i.e., 6 mm from the center of one capacitor to the center of the next capacitor). We could define our active area, that is, the area in which we can measure a change in capacitance, as aligning with the centers of the outer capacitors (Fig. 2.12).

However, we know that if a finger is outside of the last row or column, it will still have an effect on the outer capacitors it is closest to. To make things simple, let's assume we can use half the pitch around the edges, meaning that we can detect a finger up to 3 mm away from the center of our outer capacitors.

So our new active area is now defined like Fig. 2.13.

Fig. 2.12 Active area
centered on outer capacitors

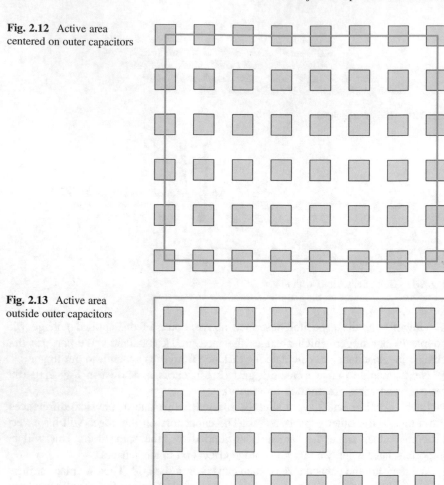

Fig. 2.13 Active area
outside outer capacitors

We need to apply a coordinate system to this area. We will follow the LCD coordinate convention since eventually we will want to compare our touch location with the location of some widget on the screen. The upper left corner is (0, 0). To keep things simple for the software engineers who prefer powers of 2, we will define the bottom right corner as (1023, 1023). To clarify later discussions involving multiple coordinate systems, we will call these coordinates touch pixels.

We have 8 rows and 6 columns, each with a pitch of 6 mm for a total active area of 48 mm by 36 mm. Those lengths are mapped into our 1024 × 1024 coordinate system (Fig. 2.14).

Fig. 2.14 Active area dimensions

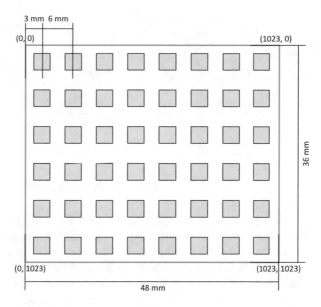

Fig. 2.15 Locations of rows and columns

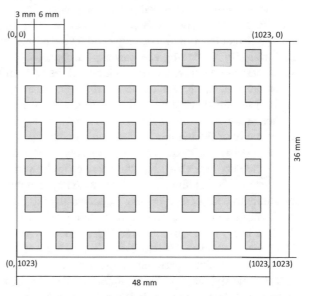

In the X direction, each touch pixel is $(48 / 1024) = 0.047$ mm. In the Y direction, each touch pixel is $(36 / 1024) = 0.035$ mm. Since the center of the first column is 3 mm from the edge, the X coordinate for the center of the first column is $((3 / 48) * 1024) = 64$ touch pixels. Applying this idea to each column and row gives the coordinates for the edges and the centers of each column and row shown in Fig. 2.15.

If a finger is 72% of the distance between columns 5 and 6 and 31% of the distance between rows 3 and 4, then the finger location in touch pixels is

$$x = \left(\frac{3\ \text{mm} + 6\ \text{mm} * \text{Col} + 6\ \text{mm} * \text{Percentage}}{48}\right) * 1024$$

$$= \left(\frac{3 + 6*5 + 6*0.72}{48}\right) * 1024 = 796 \tag{2.12}$$

$$Y = \left(\frac{3\ \text{mm} + 6\ \text{mm} * \text{Row} + 6\ \text{mm} * \text{Percentage}}{36}\right) * 1024$$

$$= \left(\frac{3 + 6*3 + 6*0.31}{36}\right) * 1024 = 650 \tag{2.13}$$

$$\text{Coordinates} = (796, 650) \tag{2.14}$$

We now know how to use a grid of capacitors to locate a touch within that grid. The touch drains some of the energy away from the closest capacitors. By comparing the new measured capacitor values with our original, baseline capacitance measurements, we can locate a touch as being between two specific columns and two specific rows. And with a little interpolation, we can narrow down that position between the columns and rows, resulting in a very accurate position measurement. We can then translate that position to touch pixels using whatever scale we want.

So far we have only discussed one finger. What if we put two fingers on the grid in two different locations? The fingers affect the nearest capacitors in their respective grid areas. By measuring the changes in capacitance, we can easily detect both fingers at the same time. It does not matter if the two fingers are exactly on the same row or column. Since each capacitor in the grid is a separate sensor, each capacitor can be used independently to locate touches. In theory, we can place as many touches as we want on the grid and detect all of them. In practice, there are a few things that limit the number of detectable touches.

The first issue with detecting two touches is how close together they are. If the fingers are very close together, the changes in the surrounding capacitors will look the exact same as if there was a single, large finger in the same area. We can improve our ability to detect close fingers by decreasing the pitch of the rows and columns, but of course there are practical limits to how close they can get. There is also the physical limit of the size of the fingers themselves. An average adult finger is 10–12 mm in diameter. If a finger is 10 mm wide at its thinnest cross section, then the maximum distance between two adjacent fingers is also 10 mm. But what about a stylus? Some styli are 1–2 mm in diameter. It is very difficult to detect two styli that are very close together (actually it is difficult to detect a 1 mm stylus in the first place, but we will address that in the chapter on stylus and glove touches).

The second practical limit to the number of touches that can be detected is the processing time it takes to determine the position of each touch. We need to have a reasonable report rate for each touch. If we are spending 100 ms processing 100 separate touches, we are going to see very jerky movements in our touch reporting. Because of these two practical limits, all touch controllers restrict the number of reported touches. Some controllers, generally targeted at smaller screens or wearables, are limited to four touches. Other controllers support 10–20 touches. We might wonder why a controller would ever support more than 10? There are large touch screen applications where multiple people are using the touch sensor at the same time. Imagine, for example, a 32 in touch surface with four people playing a touch-enabled form of doubles tennis. Fortunately, since our grid of capacitors allows us to determine the location of multiple touches, we can use it for these kinds of multi-touch applications.

Up to this point we have only discussed what is known as mutual capacitance. Mutual capacitance refers to a sensor where each capacitor has a positive and negative plate. In other words, each one is a complete capacitor. But what if we only have one plate of the capacitor? Can we even make a capacitor out of just one plate? In fact we can! This kind of capacitance is called self capacitance and is dealt with in the next chapter.

Chapter 3
Self Capacitance

In the first chapter we explained how a capacitor is formed by placing two parallel conductive plates very close to each other and applying a voltage to one plate (Fig. 3.1).

What if we only have the top plate (Fig. 3.2)?

Does this even form a capacitor? You might be surprised to find that it does! The second plate of the capacitor is essentially a virtual ground created by the environment around the plate. You can think of it as a grounded sphere surrounding the plate (Fig. 3.3).

The charge in the plate slowly leaks out to the surrounding environment. This by itself is not very interesting since the plate loses charge very slowly and therefore reaches the charging voltage very quickly. But if we put a grounded rod near the plate (Fig. 3.4), then the plate charges up much more slowly since a lot more of the charge goes into the grounded rod.

And just like we did with the two plate capacitor in the previous chapter, we can measure the difference in charge time between the first case (no rod) and the second case (rod) and use the time difference to calculate the distance to the rod. And just as before, if we replace the rod with a finger which looks like virtual ground, then we can calculate the distance to the finger.

Because this method uses a single plate capacitor, it is called self capacitance. The main difference between self capacitance and mutual capacitance is the size of the electric field. With mutual capacitance we get a very small electric field. The field exists almost entirely between the plates with some residual field, called a fringe field, around the edges (Fig. 3.5).

With self capacitance, the field stretches out into space essentially looking for something to couple to (Fig. 3.6).

Because the self capacitance field is not confined between two plates, it is much larger. That makes it much easier to detect grounded objects further away. That is why self capacitance is often used for things like proximity sensing where we want to detect an object that is several centimeters away from the sensor plate. It is also

© Springer International Publishing AG, part of Springer Nature 2019
T. Gray, *Projected Capacitive Touch*, https://doi.org/10.1007/978-3-319-98392-9_3

Fig. 3.1 Two plate
capacitor

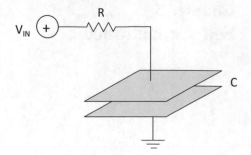

Fig. 3.2 Single plate
capacitor

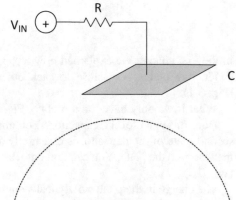

Fig. 3.3 Single plate in a
grounded sphere

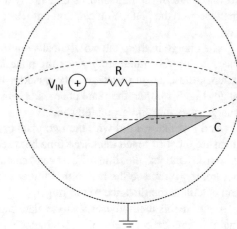

Fig. 3.4 Single
plate and
grounded rod

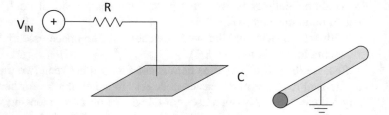

Fig. 3.5 Two plate electric
field

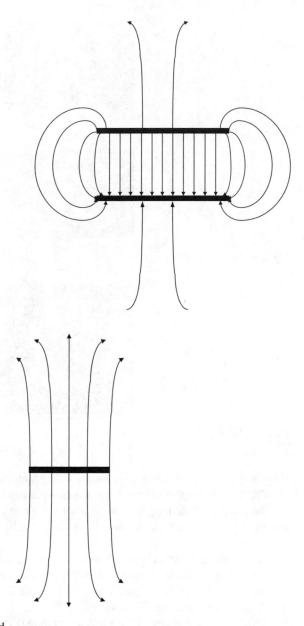

Fig. 3.6 Single plate electric field

used for applications like sensing the level of a liquid in a tank where the electric
field couples to the liquid.

The most common use for self capacitance in user interfaces is for buttons,
sliders, and wheels, often referred to by the acronym BSW. A common example
of a capacitive button is the collection of touch buttons on a cell phone. These
nonmechanical buttons are often designed using self capacitance because of the

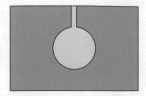

Fig. 3.7 Button pad

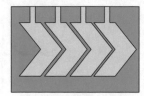

Fig. 3.8 Slider pads

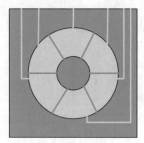

Fig. 3.9 Wheel pads

simplicity of the design. Implementing a self capacitance button can be as simple as creating a round copper electrode on a printed circuit board (Fig. 3.7).

The pad serves as the positive plate of the electrode. Connect it to a charging circuit, measure how long it takes to reach a certain voltage, and you have got a capacitive button.

Sliders are just collections of consecutive capacitive plates. They are often designed as chevrons to ensure a smooth transition from plate to plate. With some complicated interpolation algorithms, it is possible to get surprisingly high resolution from a small number of plates (Fig. 3.8).

A wheel is just a slider in the shape of a circle (Fig. 3.9).

The electrode designs shown are a bit simplistic and can have some problems with noise and unintentional touch activations. The actual electrode design depends on a number of factors including the capacitive controller being used, the sensor material, the number of available layers, the distance between the sensor and the controller, the enclosure design, and a large number of other factors. One complicating factor is backlighting. Some designs require a backlight in the center of the button or wheel. Typically this is accomplished by drilling a small hole through the

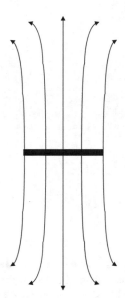

Fig. 3.10 Single plate electric field

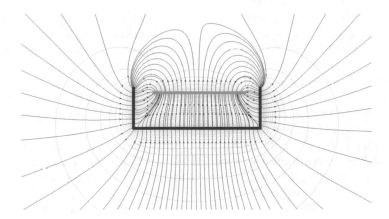

Fig. 3.11 Electric field with grounded enclosure

PCB in the center of capacitive element and reverse mounting a surface mount LED on the opposite side of the board so that the LED faces into the hole. Another option is to use a transparent conductor for the capacitive element and place lighting behind the transparent material.

You may have noticed that the electrical field lines of a self capacitance plate extend in both directions away from the plate (Fig. 3.10).

In other words, the electric field is stretching out in both directions looking for something to couple to, including other components in the system and the enclosure itself. This can alter the baseline capacitance and make it more difficult to detect a

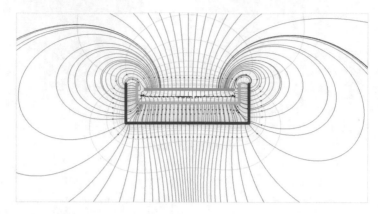

Fig. 3.12 Electric field with guard trace

finger approaching the sensor. In some self capacitance systems, an active guard trace is used to isolate the sensor from the rest of the system. Figure 3.11 shows a simulation of the electric field when the sensor is surrounded by a grounded enclosure. The flat plate in the center is the sensor. The U-shaped plate underneath and on the sides is the enclosure.

The electric field of the capacitor is very heavily coupled to the enclosure to the point that just placing a finger on the enclosure could alter the capacitance enough to indicate a touch. Also notice that the electric field does not extend very far toward the user because a lot of it is being pulled down to the enclosure. If we add an active guard trace between the enclosure and the sensor, which is basically a trace with a driven positive voltage on, then the electric field is redirected to extend farther up toward the user (Fig. 3.12). The guard trace is the second U-shaped element between the sensor and the enclosure.

As a result, the change in capacitance is much greater when a finger is present. And touching the enclosure does not affect the capacitor. The active guard trace also has the advantage of shielding the sensor from radiated noise in the system that might interfere with measuring the sensor.

Another major difference between mutual and self capacitance is the measuring technique. With mutual capacitance there are two plates, positive and negative. Typically the touch controller drives one plate and measures another. But with self capacitance there is only one plate. So the controller has to drive the plate *and* measure the capacitance on the same controller pin. There are several techniques for doing this including capacitive voltage divider (CVD), sigma delta modulator (CSD), relaxation oscillator method (ROM), and others. Most companies that sell BSW controllers provide excellent documentation on their particular measurement method. A number of the BSW vendors also have design guides with detailed information on electrode design, guard traces, noise avoidance, PCB layout guidelines, etc.

A word of advice on the use of a self capacitance controller: While it is possible to roll your own self capacitance algorithm on a microcontroller, as with many technologies, the devil is in the details. You might develop an application that works well in the lab under controlled conditions but is plagued by issues in the field caused by noise, varying ground schemes, and even different users. The companies that develop and sell BSW controllers have put a lot of thought, analysis, and testing into their solutions and their toolsets. Using an off-the-shelf BSW controller can save a lot of development time and mitigate the risk of potential field failures and returns. In other words, why reinvent buttons, sliders, and wheels?

Chapter 4
Measuring Capacitance

In a previous chapter we discussed how a mutual capacitance touch screen is essentially a grid of capacitors (Fig. 4.1).

But how do we actually connect to those capacitors? We could run separate wires for each plate, but that creates a LOT of connections (Fig. 4.2).

We are only showing connections for the first row, and already the number of connections is difficult to track. And our simple sensor only has 48 capacitors. Larger touch screens can have several thousand. Instead of running a separate wire for each capacitor, let's put the top plates for each row on one connection and the bottom plates for each row on another connection (Fig. 4.3).

Now we have two connections per row, and all the capacitors in each row are in parallel. That helps with the connections, but it also takes away the resolution. If all the capacitors in the first row are in parallel, then we can only measure the capacitance of the entire row, not of an individual capacitor. We will be able to tell if a finger is near some of the capacitors on that row, but we will not be able to tell which specific capacitors are being affected.

At this point, we borrow a concept from keyboard design. When designing a circuit to monitor an array of keys, it is common to arrange the keys into a matrix of rows and columns, and then have a connection for each row and each column. When a key gets pressed, it shorts a particular row to a particular column. By driving the rows one at a time and measuring the columns, we can determine exactly which key is pressed. We can do the same thing for our grid of capacitors (Fig. 4.4).

It may look odd to connect capacitors in this way, but it works! Once we add the connections for all the capacitors, we end up with one connection for each row and one for each column. In our simple example, that is $8 + 6 = 14$ connections (Fig. 4.5).

These rows and columns are called electrodes (for reasons that will be a bit more obvious when we describe how the capacitors are physically created), so this touch sensor requires 14 electrodes. Since we will eventually be getting coordinates out of this sensor, we can think of the rows as the X coordinates and the columns as the Y coordinates and rename our electrodes to X and Y. Since coordinate systems start

© Springer International Publishing AG, part of Springer Nature 2019
T. Gray, *Projected Capacitive Touch*, https://doi.org/10.1007/978-3-319-98392-9_4

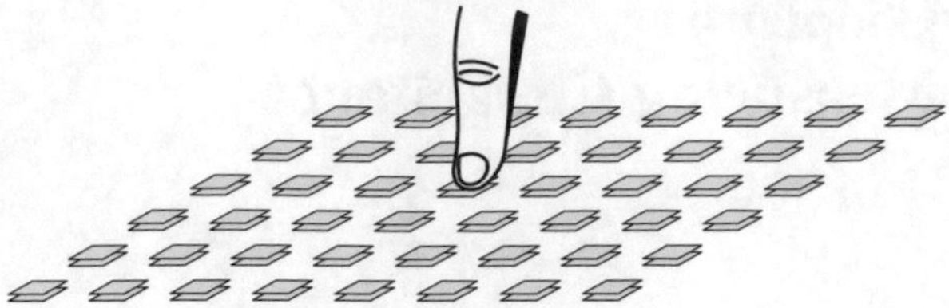

Fig. 4.1 Grid of capacitors
with finger

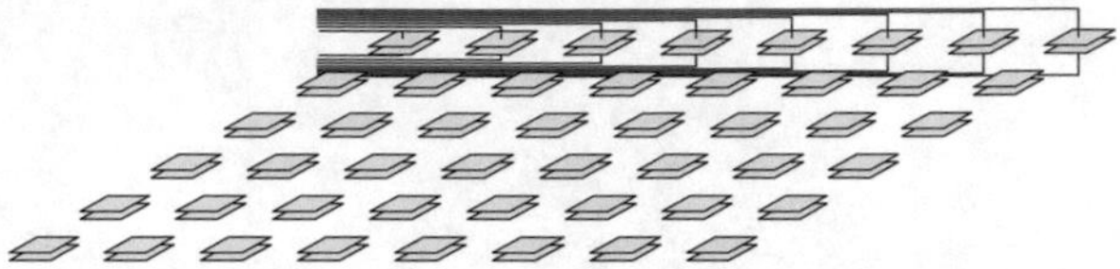

Fig. 4.2 Individual
capacitor connections

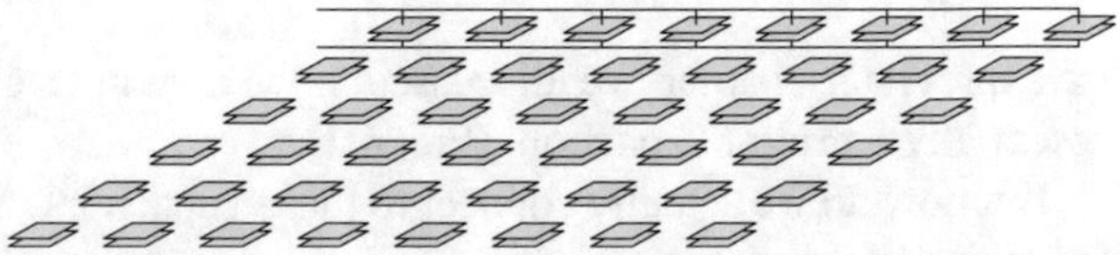

Fig. 4.3 Row capacitor
connection

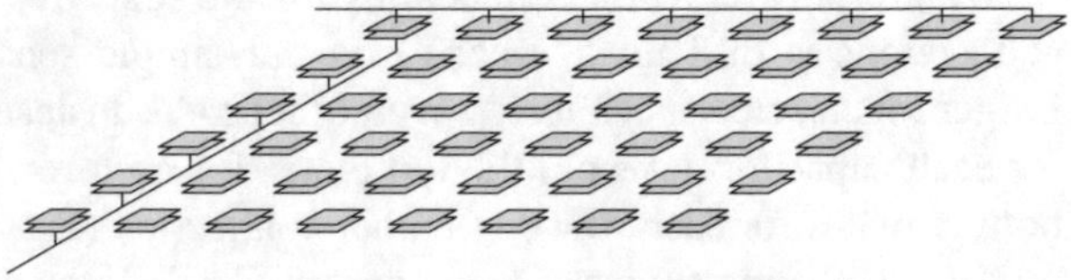

Fig. 4.4 Row and column
capacitor connections

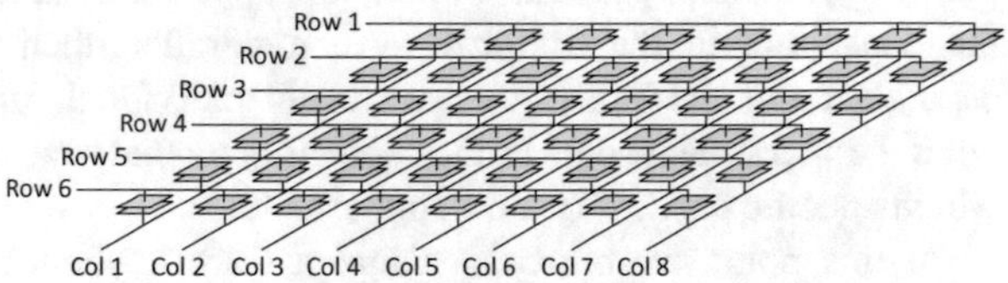

Fig. 4.5 All row and
column connections

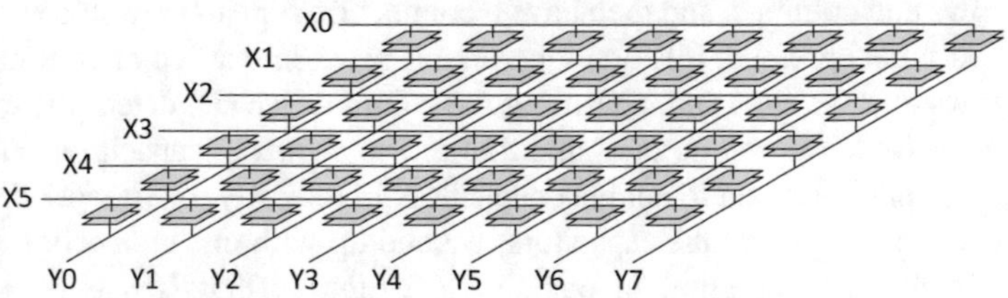

Fig. 4.6 X and
Y connections

with 0, and since the software engineers who write the code to manage all this consider 0 a perfectly valid number, the electrodes typically start at X0 and Y0 (Fig. 4.6).

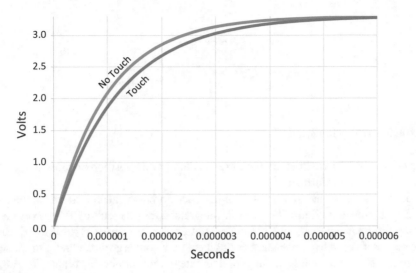

Fig. 4.7 Capacitor charging

To measure the capacitance of a particular capacitor (which is called a node), we drive one of the X electrodes and measure all of the Y electrodes. Ideally we would measure all the Y electrodes simultaneously, but that requires a lot of analog circuitry for every Y electrode. Some touch controllers multiplex the Y electrodes so that a subset (typically four or eight electrodes) is measured at one time. After the first group of Y electrodes are measured, the controller switches the multiplexer to the next group of Y electrodes and drives the same X electrode again. This is repeated until all Ys have been measured. Obviously this takes longer which reduces the number of total scan cycles per second, which is why it is less than ideal.

Getting back to the measurement, an X electrode is driven high and the corresponding voltage on a given Y electrode is measured. Remember from an earlier chapter that the voltage on Y after a certain amount of time can be used to determine the capacitance of that node. Placing a finger near the node changes the capacitance which changes the charge curve (Fig. 4.7).

We said before that we can determine the capacitance by measuring the amount of time it takes to charge up to a certain voltage. But if you look at the graph in Fig. 4.7 more closely, you will see that this creates some difficulties. The two curves are very close to each other for the first part of the graph. They diverge a bit in the middle section but then get close to each other again near the end. If we measure the voltage near the beginning or near the end, we would have a difficult time telling the difference between the two voltages. Of course the obvious choice would be to measure somewhere in the middle where the divergence of the two curves is maximum. But figuring out exactly when to take the measurement would be difficult. And the timing of the measurement would change slightly depending on the original capacitance and the new capacitance. Looking at it another way, the information we want (i.e., the difference between the two curves) is spread out over

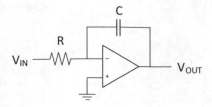

Fig. 4.8 Y voltage integration

the entire curve. If we limit our measurement to just one position on those curves, we lose a lot of that information.

Just looking at the curve and thinking back to basic calculus for a minute, the answer becomes obvious. We need to integrate across the curve! That gives us *all* of the information about the difference between the two curves regardless of where the curves diverge from one another or where they converge at the end. Fortunately, there is a standard analog circuit called an integrator that does just that (Fig. 4.8).

$$V_{OUT} = \int_0^t -\frac{V_{IN}}{RC} dt \qquad (4.1)$$

The accumulated voltage on the integrator represents the capacitance. In theory, the controller could convert the integration measurement into an actual capacitance value, but there is really no need. What we are interested in is how much the capacitance changes. All we have to do is measure the integrated voltage without a finger present (which becomes our baseline measurement). Then we continually monitor the integrated voltage to see how much it has changed from the original baseline. If it changes by more than some amount, then a touch is present (we will get into the specifics of how a touch is qualified later). The key point is that we are interested in a ratiometric measurement, so the units of the actual measurements do not matter.

Figures 4.9 and 4.10 show what the voltages on the X and Y electrodes look like for an actual touch sensor. The X (drive) voltage is on channel 1 in yellow. Channel 2 shows the corresponding Y (sense) voltage in blue. Fig. 4.9 shows the voltages when no touch is present.

In this case, the X pulse is about 8 V. The corresponding voltage on the Y electrode is around 90 mv. Fig. 4.10 shows the voltages with a touch.

On this particular sensor and controller, the change in capacitance is around 0.09 pF, or 7% of the baseline. As you can see, the difference in the Y signal is so small that it is very difficult to even see a difference between the two signals. Fig. 4.11 shows the two signals overlaid on top of each other. The touch signal is the lower of the two.

Even though it is difficult to see the difference in voltage on the scope, the difference between the two integration measurements (i.e., the difference between the voltage area under each Y pulse) will be large enough to determine that a finger is nearby. Also notice that the voltage with the touch reaches a constant voltage level that is less than the drive pulse voltage. If the node acted like a normal capacitor, the

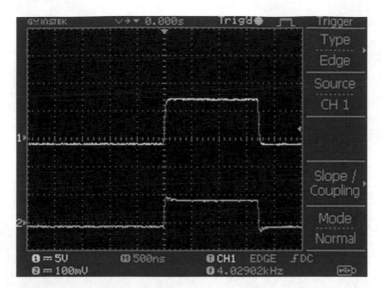

Fig. 4.9 X and Y voltages – no touch

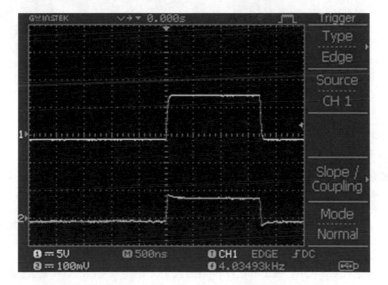

Fig. 4.10 X and Y voltages – with touch

touch and non-touch voltages would be very close to equal after the rise time. They are different because adding a finger does more than just add a single capacitance. In the next chapter we will explore the equivalent circuits with and without a touch and see how much the two circuits differ.

Keep in mind that we are driving one side of the capacitor and measuring the other side. That makes this a mutual capacitance sensor. In the previous chapter, we

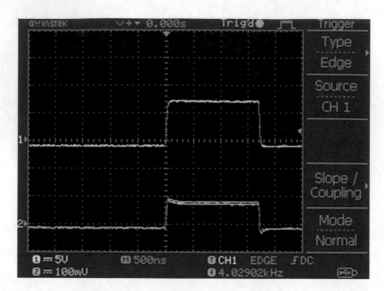

Fig. 4.11 X and Y voltages – touch and no touch

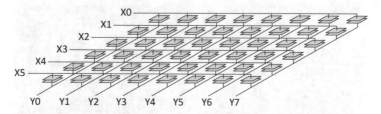

Fig. 4.12 All X and Y connections

explained how self capacitance can be used for buttons, sliders, and wheels. We can also use it for a grid touch sensor, but there are some drawbacks. Before we get to those, let's delve into how a self capacitance touch screen works.

The grid concept is the same as before with X electrodes going in one direction and Y electrodes going the opposite direction, and the tops connected in rows and the bottoms connected in columns (Fig. 4.12).

The difference is, instead of driving one X electrode at a time, we drive them all simultaneously. Then we drive all the Y electrodes simultaneously. For each electrode, we use one of the self capacitance measurement techniques mentioned in the previous chapter to measure the change in self capacitance of each electrode. Let's say as an example that we see a change in capacitance on electrodes X2 and Y3. From this we can deduce that there is a touch happening at the intersection of these two electrodes, node X2Y3.

One of the upsides of using self capacitance is that the scan cycle is much faster. Instead of driving each X electrode one at a time, you drive all the X electrodes simultaneously, then all the Ys. But this is also one of the downsides of using self

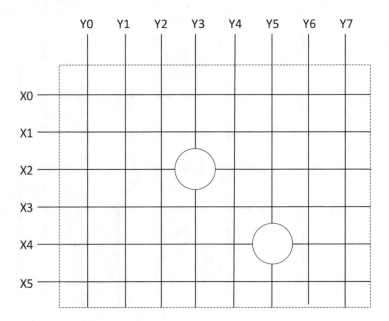

Fig. 4.13 Two touches

capacitance. Let's say I have two diagonal touches on the sensor, one at X2Y3 and another at X4Y5 (Fig. 4.13).

The controller sees a change in capacitance on X2, X4, Y3, and Y5. At first glance this may not seem like a problem. The controller can just match up X2Y3 and X4Y5. But what if the touches are actually at X4Y3 and X2Y5 (Fig. 4.14).

The controller sees a change in capacitance on the same electrodes: X2, X4, Y3, and Y5. In other words, the controller cannot tell the difference between the two sets of positions, so which positions does it pick?

While this is a potential issue, it often is not a problem in real life since it is extremely unlikely that two touches will appear simultaneously on the touch sensor between the same two scan cycles (a self cap scan cycle is typically 1–5 ms). More commonly one touch will appear during one scan cycle and the second touch will occur during a later scan cycle. With some minimal amount of software tracking, the controller can determine which electrode combinations are real touches and which are ghost touches. But as more fingers get added, the touch determination and tracking get more ambiguous. That is why self capacitance touch sensors are often limited to two touches in software. And even with sophisticated tracking algorithms, they can still get confused by complicated finger movements, especially rotations.

The reason for the difficulty of resolving multiple touches with self capacitance is that all of the electrodes are driven at once. Why not drive the electrodes one at time and get the improved resolution that is inherent in mutual capacitance? There are several reasons. First, just driving one electrode at a time does not improve the resolution. You can still only sense a change in capacitance on each electrode, not on

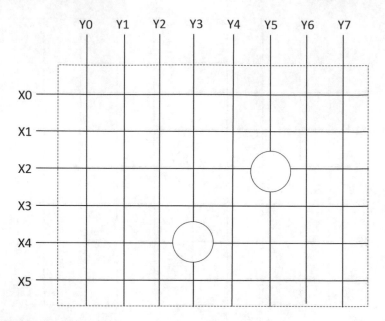

Fig. 4.14 Two touches reversed

each node. Second, if you drove one electrode at a time, say X3, then instead of measuring the self capacitance of X3 you would really be measuring the capacitance between X3 and the other surrounding electrodes including the Y electrodes running underneath the X3. The only way to prevent X3 from coupling to the electrodes around it is to drive those other electrodes with the same voltage as X3. This is very similar to the guard track idea used to increase the self capacitance of a button that was mentioned in the previous chapter. So now we are talking about driving all the electrodes at once, which is exactly what we are trying to avoid doing. So driving one electrode at a time does not help.

You might be wondering why we are talking about using self capacitance at all for a touch sensor given that it does not handle multiple touches very well and has lower resolution than mutual capacitance. Let's review the strengths and weaknesses of mutual and self capacitance for a touch screen (Table 4.1).

As you can see, the pros of self and mutual capacitance are complimentary. For this reason, some touch controllers support both types of measurements. They use mutual capacitance for high-resolution multi-touch tracking, and self capacitance for low-resolution single finger tracking. By combining these two measurement schemes, it is possible to track multiple touches under most conditions, then switch to single finger tracking mode when the user is wearing a thick glove or when water is present. By supporting both measurement types, a touch controller is able to handle a larger number of use cases.

One of the biggest benefits of combining mutual and self is that water on the sensor affects self measurements much less than it affects mutual measurements. By analyzing both sets of data, a touch controller can ignore touches that appear in the

Table 4.1 Mutual and self capacitance pros and cons

Pros	Cons
Mutual capacitance	
High resolution	Small electric field (i.e., low sensitivity)
Track a (theoretically) infinite number of touches	Long scan cycles
	Water on the sensor causes changes in capacitance
Self capacitance	
Large electric field (i.e., high sensitivity)	Low resolution
Short scan cycles	Difficult to track more than one touch
Water on the sensor has little impact on capacitance	

mutual data but do not appear in the self data. We will cover this in more detail in a later chapter.

When describing the capacitance measurements in this chapter, we have implied that each mutual or self measurement is measuring one capacitance. From the controller's point of view this is entirely accurate. The controller can only measure one value per node during a mutual scan or one value per electrode during a self scan, so it is seeing only one capacitance. In reality, the measured capacitance is actually a combination of multiple capacitances and resistances. To understand what the various capacitances are, in the next chapter we look at some equivalent circuits of a touch sensor.

Chapter 5
Equivalent Circuits

We explained in the previous chapter that in mutual capacitance sensors the top electrodes of the capacitors are connected in rows and the bottom plates are connected in columns. We have drawn the electrodes as plate capacitors connected by skinny traces (Fig. 5.1).

The implication in this picture is that each capacitor is a separate entity and could be considered as an independent component. It probably will not come as a surprise that it is not that simple! In fact, the top plate of a particular capacitor has a capacitive coupling to all of the elements around it. To simplify things, let's consider a single capacitor or node, X2Y2 (Fig. 5.2).

We have included the X electrodes and Y electrodes adjacent to X2 and Y2 and removed all the other electrodes to simplify the diagram.

In this example, we are putting a pulse on X2 and measuring Y2. The complication is that the energy from the pulse on X2 does not just get coupled to Y2. Some of that energy goes into the X1 and X3 electrodes. Some of it goes into Y1 and Y3. And some of it goes into ground. What that really means is that we have capacitors between all of those points (Fig. 5.3).

In Fig. 5.3, each capacitor is labeled by the two items that it connects. For example, capacitor X2Y3 represents the capacitance between the X2 and Y3 electrodes. Capacitor X1G represents the capacitance between the X1 electrode and ground, and so on.

This simple diagram of three X electrodes and three Y electrodes results in 19 capacitors! In theory, we could create a circuit model that includes all 19 capacitors. But in reality we do not need to model all 19 of them for several reasons.

The first step in understanding which capacitors can be ignored is to understand a bit more about how a controller takes a capacitance measurement. As we said before, the controller drives one X electrode and measures all the Y electrodes (the controller may multiplex the Ys which requires multiple pulses on the same X electrode, but we are ignoring that complication for now). When the controller is pulsing an X electrode, what is happening with the X electrodes on either side of the driven electrode? The controller grounds them. In our simplified example, X1 and X3 are

© Springer International Publishing AG, part of Springer Nature 2019
T. Gray, *Projected Capacitive Touch*, https://doi.org/10.1007/978-3-319-98392-9_5

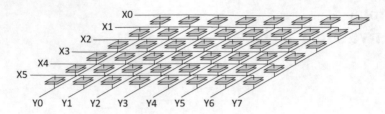

Fig. 5.1   Grid of capacitors

Fig. 5.2 3×3 grid

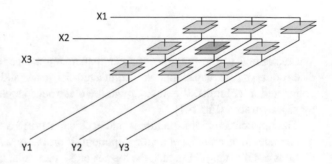

Fig. 5.3   Electrode capacitances

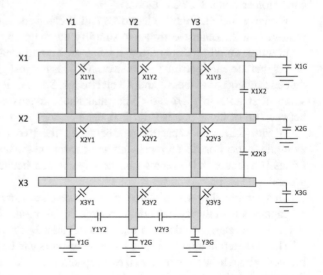

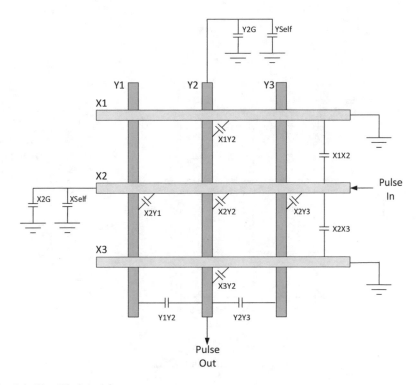

Fig. 5.4 Simplified model

grounded. So we can also eliminate the X1G and X3G capacitors because those electrodes are shorted directly to ground. And we can eliminate X2G because that electrode is being driven by a voltage source and a series capacitor just adds a little delay. We can also eliminate the capacitors between the grounded electrodes and the Y1 and Y3 electrodes since right now we are only interested in Y2. In reality, these capacitances do exist, but they are very small compared to the other capacitors in the system. Figure 5.4 shows the simplified model.

We also *added* two capacitors, XSelf and YSelf, to represent the self capacitance of the two electrodes we are interested in. Now we need to add components to represent the incoming drive pulse and the measurement of the outgoing pulse. The drive pulse is just a DC voltage source. But, as we will learn later when we talk about the materials used to construct a touch sensor, the electrode has a nontrivial resistance that we need to include in our model. The sense electrode also has resistance. For the sense measurement circuit itself, we will model it as a simple integrator. We are going to ignore any impedance in the drive circuit. Figure 5.5 shows the model with the drive and measurement circuits added.

We can convert this model into a circuit diagram by thinking of each electrode and ground as a node and connecting all the nodes together, which produces the equivalent circuit shown in Fig. 5.6.

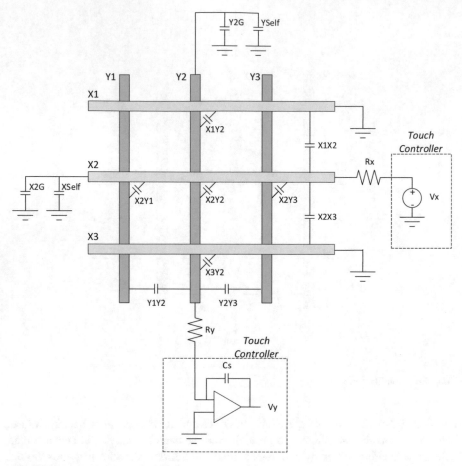

Fig. 5.5 Measurement connections

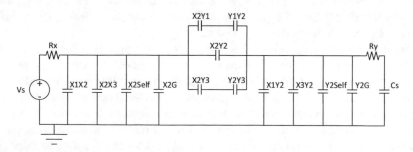

Fig. 5.6 Equivalent circuit

One of the simplifications in this circuit is that we only model one resistor for the X electrode and one for the Y electrode. A more accurate model would split each of these two resistors into two more resistors. One resistor would represent the

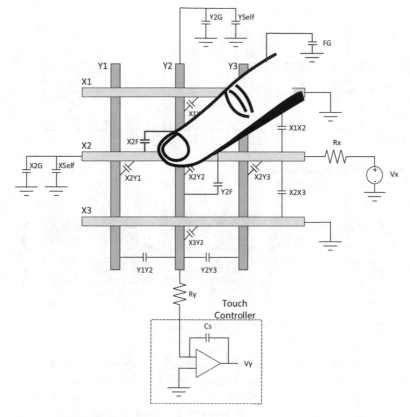

Fig. 5.7 Electrodes with finger

resistance between the controller and the node we are interested in. The second resistor would model the resistance between the node and ground. We have left these out just to keep the circuit from becoming too complicated, and because resistance after the finger (i.e., on the opposite side of the finger from the drive or sense circuit) docs not have much impact on the measurement.

What happens when a finger is on the sensor? We get three new capacitors (Fig. 5.7).

There is a capacitance between the finger and ground (FG), a capacitance between the finger and the X2 electrode (X2F), and a capacitance between the finger and the Y2 electrode (Y2F). Adding these to our equivalent circuit, we get Fig. 5.8.

If you are familiar with transfer functions, the transfer functions for the two circuits are shown in Eqs. 5.1 and 5.2.

$$H(s) = \frac{C_2}{s^2 \left(R_x R_y C_s \left(C_1 C_3 + C_1 C_2 + C_2 C_3 \right) \right) + s \left(R_x \left(C_1 C_3 + C_1 C_5 + C_1 C_2 + C_2 C_3 + C_2 C_5 \right) + R_y \left(C_3 C_5 + C_2 C_5 \right) \right) + C_2 + C_3 + C_5} \tag{5.1}$$

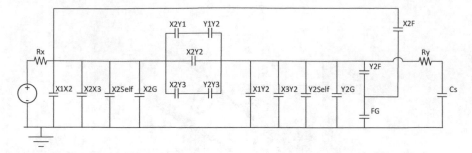

Fig. 5.8 Equivalent circuit with touch

$$H(s) = \frac{C_{FG}C_2 + C_{X2F}C_2 + C_{Y2F}C_2 + C_{Y2F}C_{X2F}}{\begin{aligned}&s^2\left(\begin{aligned}&C_SC_{FG}C_2C_1R_XR_Y + C_{X2F}C_SC_2C_1R_XR_Y + C_{Y2F}C_SC_2C_1R_XR_Y + C_SC_{FG}C_3C_1R_XR_Y + C_{X2F}C_SC_3C_1R_XR_Y + C_{Y2F}C_SC_3C_1R_XR_Y + \\&C_{Y2F}C_SC_{FG}C_1R_XR_Y + C_{Y2F}C_{X2F}C_SC_1R_XR_Y + C_SC_{FG}C_3C_2R_XR_Y + C_{X2F}C_SC_3C_2R_XR_Y + C_{Y2F}C_SC_3C_2R_XR_Y + C_{X2F}C_SC_{FG}C_2R_XR_Y + \\&C_{Y2F}C_SC_{FG}C_2R_XR_Y + C_{X2F}C_SC_{FG}C_3R_XR_Y + C_{Y2F}C_{X2F}C_SC_3R_XR_Y + C_{Y2F}C_{X2F}C_SC_{FG}R_XR_Y\end{aligned}\right)\\&+s\left(\begin{aligned}&C_{FG}C_2C_1R_X + C_{X2F}C_2C_1R_X + C_{Y2F}C_2C_1R_X + C_{FG}C_3C_1R_X + C_{X2F}C_3C_1R_X + C_{Y2F}C_3C_1R_X + C_SC_{FG}C_1R_X + C_{Y2F}C_{FG}C_1R_X + \\&C_{X2F}C_SC_1R_X + C_{Y2F}C_SC_1R_X + C_{Y2F}C_{X2F}C_1R_X + C_{FG}C_3C_2R_X + C_{X2F}C_3C_2R_X + C_{Y2F}C_3C_2R_X + C_SC_{FG}C_2R_X + C_SC_{FG}C_2R_X + \\&C_{X2F}C_{FG}C_2R_X + C_{Y2F}C_{FG}C_2R_X + C_{X2F}C_SC_2R_X + C_{X2F}C_SC_2R_X + C_{Y2F}C_SC_2R_X + C_{Y2F}C_SC_2R_X + C_SC_{FG}C_3R_Y + C_{X2F}C_{FG}C_3R_X + \\&C_{X2F}C_SC_3R_Y + C_{Y2F}C_SC_3R_Y + C_{Y2F}C_{X2F}C_3R_X + C_{X2F}C_SC_{FG}R_X + C_{Y2F}C_SC_{FG}R_Y + C_{Y2F}C_{X2F}C_{FG}R_X + C_{Y2F}C_{X2F}C_SR_Y + \\&C_{Y2F}C_{X2F}C_SR_X\end{aligned}\right)\\&\left(C_{FG}C_2 + C_{X2F}C_2 + C_{Y2F}C_2 + C_{FG}C_3 + C_{X2F}C_3 + C_{Y2F}C_3 + C_SC_{FG} + C_{Y2F}C_{FG} + C_{X2F}C_S + C_{Y2F}C_S + C_{Y2F}C_{X2F}\right)\end{aligned}} \quad (5.2)$$

We will not explore the transfer functions any further as they do not contribute much to the understanding of how a touch sensor functions. We would also have to get pretty deep into Laplace transforms, poles, and differential equations. As college textbooks like to say, "This is left as an exercise to the reader."

In an earlier chapter, we said that adding a finger to a touch sensor increases the capacitance. Comparing the two circuit models, you can see that it is not that simple. Ideally we would like to develop an equation that calculates the difference in capacitance between no finger and finger, something that would give us a simple answer like, "a typical finger increases the capacitance by 5 pF." But since the finger interacts with the node being measured, with the X electrode being driven, and with the Y electrode being measured, the effect of adding a finger is more complicated than just the addition of extra capacitance. The transfer functions are also good indicators of how complicated the calculations get when a finger is added.

But the good news is that, from the controller's standpoint, all those complications do not matter. All that matters is that the time constant of the circuit changes when a finger is present. We can use this change in time constant to determine when a finger is near. And since we are comparing a baseline measurement with no finger to a dynamic measurement with a finger, we are essentially looking at a ratio of two times. We do not even have to convert the measurement into a capacitance. We just need to know that that measured time constant increases by 7% or whatever is typical for a finger for this particular sensor.

Now that we understand how a projected capacitive sensor works, the next logical step is to physically construct a sensor. That starts with finding a conductive material that we can see through. And in order to compare the various transparent conductors, we first need to understand the optical characteristics used to qualify them. Then we need to figure out what our actual electrode patterns are going to be. Those are our next two topics.

Chapter 6
Measuring Optical Quality

The simplest way to make a touch sensor is with a printed circuit board (PCB). We can pattern the copper on the top and bottom into our capacitors, route the row and column connections to the edges, put a couple of connectors on the board, and voilà, we have a touch sensor. Of course, the downside of this is that you cannot see through it. Since most touch sensors are placed over displays, the touch sensor has to be transparent. And not just transparent, we want the display image to be as bright and clear as possible. The ideal touch sensor would have no impact on the optical quality of the display image.

Of course that is not possible in real life. The optical quality of the display image will suffer some degradation as it passes through the touch sensor layer. There are several ways to quantify this loss of quality including transmissivity, haze, gloss, and other measurements. Before we can discuss the various conductive materials used to build transparent touch sensors, we first need to understand the ways in which those materials are quantified, measured, and compared. We can then use these various metrics to compare the different transparent conductors.

In the display world, light is measured in nits. One nit is equivalent to one candela per square meter. A candela is defined in Standard Units as the amount of radiation emitted by a black body under very specific conditions. The brightness of a typical display is between 300 nits and 500 nits. High brightness displays are in the range of 800 nits to 1200 nits or more.

Transmissivity is a measure of the amount of light that passes through a transparent medium. It is the ratio of the amount of light which exits the medium to the amount of light that enters the medium. It is always less than 100%. It can also be defined as the inverse of opacity where opacity measures the amount of light that is blocked when traveling through a medium. Let's say a particular display emits 500 nits of light. We place a piece of glass over that display with a transmissivity of 90%. In this case, we could expect to measure 500 * 0.9 = 450 nits of light coming out of the glass. If we stack two pieces of the same glass on top of each other, their transmissivities multiply. So two pieces of 90% transmissivity yield 0.9 *

© Springer International Publishing AG, part of Springer Nature 2019   45
T. Gray, *Projected Capacitive Touch*, https://doi.org/10.1007/978-3-319-98392-9_6

Fig. 6.1 Photon scattering
categories

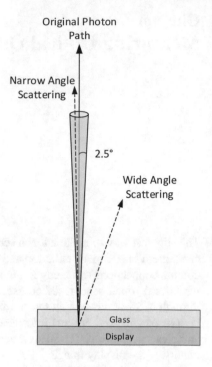

0.9 = 81% transmissivity. This will be important later when we talk about the multiple layers that make up a touch sensor.

Haze is a measurement of the divergence of a light ray from its original path. You can think of it as a more specific way of measuring transmissivity. As light passes through a transparent medium like glass, some of the photons that make up that light ray are deflected from their original paths. Some photons bounce back toward the source and never make it through the medium. This type of scattering reduces the transmissivity. Some photons are deflected from their original path but still pass through. The paths of these divergent photons can be measured as an angle of divergence from the original path. Photons that diverge from the original path by more than 2.5° are defined as having wide angle scattering. Photons that diverge by less than 2.5° are defined as having narrow angle scattering (refer to Fig. 6.1).

Haze is the amount of light that is subject to wide angle scattering. In other words, it is the amount of light that, when it emerges from the medium, has diverged by more than 2.5° from its original path. This wide angle divergence results in a noticeable haziness in the image being transmitted. Like transmissivity, haze is defined as a percentage. If a particular piece of glass has a haze of 2%, then 2% of the light passing through that piece of glass will emerge from that glass on a path that diverges by more than 2.5° from the original path.

Clarity is the percentage of light that is subject to narrow angle scattering. This is light that has diverged by less than 2.5° from its original path. When a lot of light diverges by less than 2.5°, the image suffers from a slight loss of edge definition,

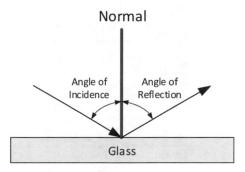

Fig. 6.2 Angle of reflection

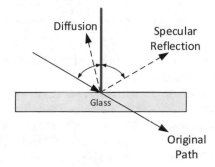

Fig. 6.3 Diffusion and specular reflection

a.k.a. clarity. Haze and clarity are obviously closely related. The datasheet for a transparent material will often include a haze spec but not clarity.

Gloss is a measure of the amount of light reflected at the surface of a material. But not just any light. When a light ray hits the surface of a transparent material, some of the light passes through, some is reflected off the surface in random directions (this is called diffusion), and some is reflected at the angle of reflection (this is called specular reflection). Refer to Figs. 6.2 and 6.3.

Gloss is the amount of light reflected specifically due to specular reflection. This can also be thought of as the mirror reflection since an ideal mirror reflects all of its light as specular reflection.

Gloss is measured in gloss units. The definition of a gloss unit is a bit confusing. It starts with a reference to a standard black glass material. The amount of specular reflection exhibited by this standard glass is defined as 100 gloss units (GU). This is measured at a 60° angle of incident for most materials (highly reflective materials are measured at 20°). The gloss of the material being tested is then normalized to this scale. If the test material is much less reflective than the standard, then it will have a gloss measurement much less than 100 GU. If it is highly reflective, it could have a very high gloss measurement. For example, mirrors can have a gloss level as high as 2000 GU.

Displays with a higher gloss rating will have more vivid color and better contrast, but they will also have more visible reflections, especially in direct sunlight. The

Table 6.1 Speed of light

Medium	Light speed (kps)
Vacuum	299,792
Air	299,000
Water	225,000
Glass	200,000
Diamond	124,000

Table 6.2 Index of refraction

Medium	Light speed (kph)	Index of refraction
Vacuum	299,792	1.0
Air	299,000	1.0003
Water	225,000	1.33
Glass	200,000	1.5
Diamond	124,000	2.42

desired gloss rating depends on the environment the display will be used in and the desired aesthetics of the product.

Index of refraction is another key measurement, although it typically matters more to the touch sensor manufacturer than the end user. It is related to the speed of light through a transparent material. The speed of light in a vacuum is the light speed most engineers are familiar with, approximately 300,000 kilometers per second (kps). You may remember from your college physics classes that light slows down when it passes through other mediums, but you might be surprised at how much slower it travels. Table 6.1 lists the speed of light through various mediums.

The ratio of the speed of light through a vacuum to the speed of light through a material is called the index of refraction. The indices of refraction have been added in Table 6.2.

When light enters a new material and its speed changes, it also changes direction. These changes in speed and direction are caused by complicated quantum wave interactions between the light wave and the atoms in the material. As the light wave enters the material, its electric field begins to interact with the electrons and protons in the material's atoms. The material's charges begin to oscillate at the same frequency as the light wave, but with a slight phase delay. This oscillation creates additional electric fields that interact with the original electric field of the light wave. The resulting light wave is the superposition of these interacting fields. This wave typically has the same frequency as the original light wave but a shorter wavelength. Explaining why it also changes direction requires some complicated math. The short version is that the light has to change direction in order to conserve momentum.

Index of refraction is important in touch sensor design because some conductive materials used in touch sensors are transparent, the most common example being Indium tin oxide or ITO. We will learn a lot more about ITO and other transparent conductors in the next chapter. The key issue for now is that the conductive material has an index of refraction just like any other transparent material. If the conductor's index of refraction does not match the surrounding material, typically glass or

Fig. 6.4 Birefringent effect

plastic, then the conductive material may be visible. To help minimize the visibility of the conductor, a process called index matching is often used so that the index of refraction of the conductive material is as close as possible to the index of refraction of the surrounding material. Again, the most common example is index matched ITO.

Birefringence is an important optical quality issue for outdoor applications. Technically, birefringence refers to transparent materials that have two indices of refraction. But when most people use the term "birefringence" in relation to touch sensors, they are referring to a rainbow-like pattern that can occur under certain conditions when looking at a display with a touch sensor (refer to Fig. 6.4). We will refer to this as the birefringent effect to differentiate it from birefringent materials.

To understand why birefringence occurs, we first have to understand some basics about display construction and about a material called a polarizer. Most displays used in the kinds of devices that have touch sensors are liquid crystal displays or LCDs. The liquid crystals are controlled by transistors, which is why another common term for these displays is TFT which stands for thin film transistor. There are several layers in a typical TFT. For now we are only going to show the layers that are important for creating the birefringent effect. A more complete explanation of TFT operation is provided in a later chapter.

For our current discussion, the key components in a TFT are (Fig. 6.5):

1. Backlight
2. Back polarizer
3. Liquid crystal
4. Front polarizer

Most modern display backlights use white LEDs. The light produced by these LEDs is unpolarized, meaning that the light waves coming from the LED are at random angles. To simplify the explanation, we will only consider a single light

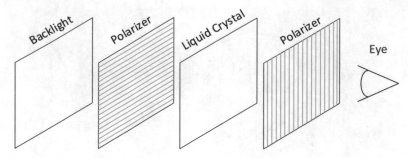

Fig. 6.5 TFT components

Fig. 6.6 Light waves at random angles

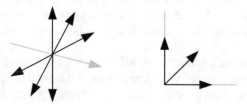

Fig. 6.7 Light wave vectors

wave coming straight out of the LED. This light wave could be at any angle around the axis of travel as shown in Fig. 6.6.

When discussing light wave orientation, it is easier to view it as a directional vector instead of a sine wave. The vector is drawn from the center of the wave to the top of the wave's peak. Figure 6.7 shows the vector representation of the same three light waves as Fig. 6.6 in both a perspective view and a straight on view (note that the negative vectors are not shown in the straight on view).

When light waves pass through a polarizer, the emerging light waves are oriented to the polarizer's direction. A common misconception about polarizers is that they only allow through light waves that are perfectly aligned with the polarizer. That is not exactly true. Light waves that are perfectly perpendicular to the polarizer are 100% blocked. Light waves that are perfectly aligned with the polarizer are 100% passed through. Light waves whose angles are not perfectly perpendicular or aligned to the polarizer have some of their intensity passed through. The amount of light passed through can be calculated as the vector projection of the light wave's vector to the polarizer's vector.

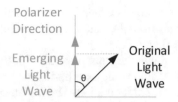

Fig. 6.8 Nonoriented light wave through a polarizer

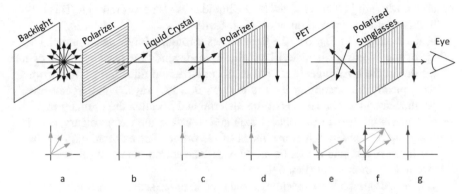

Fig. 6.9 Birefringent effect

Figure 6.8 shows what happens to a light wave that is not perfectly aligned with the polarizer. The black line represents the original light wave at some angle Θ to the polarizer. The green line represents the orientation of the polarizer. The blue arrow is the portion of the original light wave that emerges from the polarizer.

The emerging light is oriented to the polarizer and its intensity (i) is reduced to:

$$i_{new} = i_{original} * \sin(\theta)$$

If the angle is 0° (i.e., the light wave is aligned with the polarizer), then all of the light passes through ($i_{new} = i_{original}$). If the angle is 90° (i.e., the light wave is perpendicular to the polarizer), then none of the light passes through ($i_{new} = 0$). For any angle between 0° and 90°, some of the light passes through. But regardless of the original angle, any light that emerges from the polarizer is aligned with the polarizer's direction.

Figure 6.9 shows the path of a light ray starting at the backlight and finally arriving at the eye.

When light first emerges from the backlight, it is unpolarized (Fig. 6.9a). It then passes through the back polarizer and emerges aligned to the polarizer (Fig. 6.9b). Next it passes through a layer of liquid crystals. Liquid crystals can be thought of as light tubes. The tube can either be straight, in which case the light wave passes through unchanged, or it can be twisted, in which case the light wave's orientation is rotated. The liquid crystals used in TFTs are designed to have a 90° or half twist in their relaxed state. When a voltage is applied to the crystal, it untwists to an angle of 0°. So the light that emerges from the liquid crystals is either aligned with the back polarizer or is rotated 90° (Fig. 6.9c).

The light then passes through the front polarizer. This polarizer is perpendicular to the back polarizer. The diagram shows a horizontal polarizer in the back and a vertical polarizer in the front, but it could also be the other way around. The key is that the polarizers are perpendicular to each other. Two polarizers that are perpendicular to each other are called crossed polarizers.

Light that passes through an untwisted liquid crystal is blocked by the front polarizer, while light that is twisted by the liquid crystals passes through. This is how individual pixels are turned on or off (Fig. 6.9d).

In the next step in Fig. 6.9, the light passes through a birefringent material. The effect of the birefringent material on the light depends on the alignment of the molecules in the birefringent material to the polarization direction of the incoming light. Typically the incoming light is divided into two rays. The first ray travels perpendicular to the optical axis of the material and is called the ordinary ray. The second ray is aligned to the optical axis and is called the extraordinary ray. The ordinary ray passes through at one index of refraction. The extraordinary ray has a higher index of refraction so it travels slower. The extraordinary ray is also rotated 90° from the ordinary ray (Fig. 6.9e).

The most common birefringent material used in touch sensor construction is PET. Figure 6.10 shows the typical indices of refraction for the ordinary and extraordinary rays in PET.

Notice that the indices of refraction are wavelength dependent. Light with a shorter wavelength has a higher index of refraction than light with a longer wavelength. In other words, blue light (500 nm) takes longer to get through the sample than red light (650 nm). So not only is the incoming light split into two rays that are perpendicular to each other, but within each ray a slight phase delay is added based on the wavelength (Fig. 6.11).

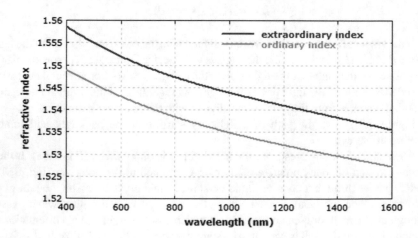

Fig. 6.10   Indices of refraction in PET. (Source: https://www.rp-photonics.com/birefringence.html)

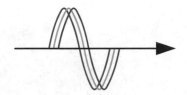

Fig. 6.11 Light retardation by wavelength

If you looked at the light coming out of the birefringent material, you would not notice anything unusual. The human eye cannot see light polarization, so the fact that the light is split into two perpendicular polarized waves would not be detectable. And the human eye cannot see slight phase delays between wavelengths, so the phase shift in colors would not be detectable. But what if the user is wearing polarized sunglasses?

When the light waves from the birefringent material hit the polarizer in the sunglasses, the two perpendicular waves are combined into a single wave. The intensity of the resultant wave can be determined by doing a vector combination of the two incoming light rays (Fig. 6.9f). But what is more important is that the light from the two waves undergoes a complex combination of constructive and destructive interference. And because the different colors in the two light waves have different phase delays, the interference creates new colors that did not exist in the original image. So what emerges from the polarized sunglasses is a polarized light wave that contains the original image as well as streaks of other colors (Fig. 6.9g). That is what we have been calling the birefringent effect.

In order to produce the birefringent effect, the birefringent material must be placed between crossed polarizers. In Fig. 6.9 the crossed polarizers are the front polarizer on the TFT and the polarized sunglasses (the back polarizer plays an important role in ensuring that only certain pixels are visible, but it does not play any role in birefringence). That is why birefringence is only a concern in outdoor applications where the user may be wearing polarized sunglasses. As mentioned above, PET is often used in touch sensor constructions, either as a substrate for the conductive material or as a substrate for an adhesive between layers. If the product is going to be used outdoors, it is important that no PET (or any other birefringent material) is used in the touch sensor construction.

Sparkle is the visual perception of light and dark areas in the displayed image (Fig. 6.12).

The intensity variation changes with the viewing angle. As the user moves in relation to the display, the display appears to sparkle. Sparkle is a common side effect of antiglare (AG) films and coatings. We will discuss AG in more detail in the chapter on cover lens enhancements. The basic idea of AG is to create a textured surface that diffuses incoming light and reduces glare. This diffusion can also create sparkle.

Fig. 6.12 Sparkle

The amount of sparkle depends on several factors including the pixel design of the TFT. Because it depends on multiple physical parameters and changes with ambient lighting and viewing angle, sparkle is subjective and difficult to measure. There is no industry standard for specifying the level of sparkle.

These various optical quality metrics will be useful in the next chapter where we compare different types of transparent conductive materials used to manufacture touch sensors.

Chapter 7
Transparent Conductors

We need some kind of conductive material to form our capacitors. But we also want to place our touch sensor over a display, which means we need to be able to see through the capacitive structures. Fortunately, there are several types of materials we can use for this. In some cases, the conductor itself is transparent. In other cases, the conductor is opaque, but extremely small so it is not visible to the naked eye. But in all cases, the conductive material needs to be placed on some kind of substrate, essentially a carrier for the conductive material. The most common substrates are glass and plastic, typically PET. The conductive material is populated onto the substrate by sputtering, roll coating, spin coating, or even inkjet printing. In most cases, the conductive material is deposited as a consistent sheet across the entire substrate. Later it is patterned into electrodes using some kind of removal process. But there are some conductors that can be deposited in the required shapes so that no additional processing is required to form the capacitors. Patterning will be discussed in the next chapter.

Before we discuss the materials, we first need to address the measurement of sheet resistivity. When a conductive material is deposited onto a substrate in a homogenous layer, the resistance of that material is specified in Ohms per square ($\Omega/\square$). What this means is that a square of the material has that resistance regardless of the dimensions. Let's say we have a sheet of conductive coated glass with a resistance of $100\ \Omega/\square$. Out of that sheet we cut a square that is 1 m by 1 m and another square that is 1 mm by 1 mm. We then measure the resistance across both squares. In both cases we will measure $100\ \Omega$.

Of course as with all things, it is not quite that simple. First, if we are actually going to measure the resistance, we need to have a low resistance connection along opposite edges of the squares. For example, we might place a strip of conductive copper tape along two opposing edges. This is required because we want the current to be evenly distributed across the entire square. Second, even if the material is exactly $100.0000\ \Omega/\square$, even if the copper tape were infinitesimally narrow (i.e., it did not encroach into the square on either edge), and even if the measurement equipment accuracy was perfect, we still would not measure exactly $100.0000\ \Omega$

© Springer International Publishing AG, part of Springer Nature 2019
T. Gray, *Projected Capacitive Touch*, https://doi.org/10.1007/978-3-319-98392-9_7

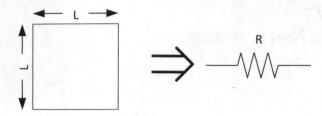

Fig. 7.1 One square

Fig. 7.2 Two squares left to right

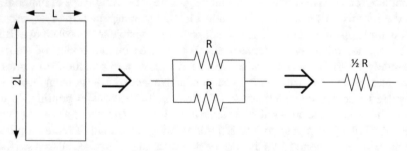

Fig. 7.3 Two squares top to bottom

on both samples. Electrons tend to travel in a nonuniform way. They prefer to travel on the surface of a material rather than through it. They also tend to travel more along the edges than the center. For these reasons, our two samples would not measure exactly $100.0000\ \Omega$. But they would be close enough that for our purposes we can consider the resistance to be $100\ \Omega/\square$ regardless of the size of the square.

So we can empirically prove that two different sample squares of the same material have the same resistance, but what is really going on? The easiest way to think of this is in terms of an equivalent resistance circuit. Let's say that we have a square of some conductive coated material. Each side of the square is of length L, and the resistance of the square is R. We can model that square of material as a resistor with resistance R (Fig. 7.1).

Now imagine that we have a rectangle of the material that is twice as long as it is high and we measure from the left edge to the right edge. We can model this as two resistors in series with a total resistance of $2R$ (Fig. 7.2).

Next we consider a rectangle that is twice as high as it is wide. This is the equivalent of two resistors in parallel which, when measured from left to right, gives a resistance of $\frac{1}{2}\,R$ (Fig. 7.3).

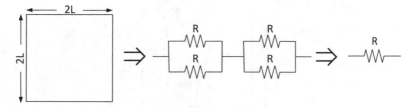

Fig. 7.4 Larger square

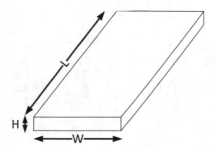

Fig. 7.5 Resistance calculation

Finally, let's put four of the squares together to form a new square twice as large as the original. The equivalent circuit for this is two groups of parallel resistors in series (or two groups of series resistors in parallel – we get the same answer either way). The two parallel resistors simplify into resistors of ½ R. Two of those in series is R (Fig. 7.4).

Therefore, given a square of conductive material of resistance R, the resistance of any size square of that material is still R.

Another way to look at it is that the resistance of a piece of material is defined by its length and cross-sectional area (Fig. 7.5 and Eq. 7.1).

$$R = \rho * \frac{L}{W * H} \tag{7.1}$$

The resistivity constant, ρ (Rho), defines how resistant a particular material is to carrying electrical charges. In a conductor like silver ρ is around 10^{-8} Ωm. In an insulator like glass ρ is around 10^{10} Ωm. Resistivity is temperature dependent. For a given material at a given temperature, ρ is constant.

Now let's assume we have a sheet of material with some resistivity, ρ. The material is a square ($L = W$) with some height H. The resistance of that material is

$$R = \rho * \frac{L}{L * H} \tag{7.2}$$

Now we double the size of the square, meaning that each side is now 2 L:

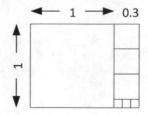

Fig. 7.6 Rectangular sheet resistance

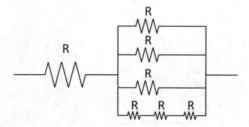

Fig. 7.7 Rectangular equivalent circuit

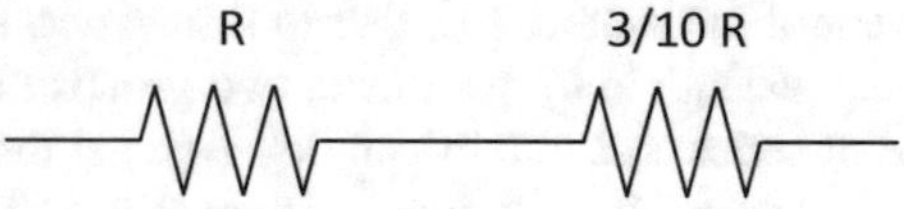

Fig. 7.8 Simplified circuit

$$R = \rho * \frac{2*L}{2*L*H} \tag{7.3}$$

As Eq. 7.3 shows, doubling the length and doubling the width cancel out, giving the same resistance value as before. In fact, Eq. 7.2 shows that if the length and width are equal, they cancel each other out so that the resistance only depends on the height. So again, any square of the material has the same resistance regardless of its dimensions. If we make a rectangle that is twice as long but the same width, then the resistance is doubled (equivalent to two resistors in series). If we make a rectangle that is half as long but the same width, then the resistance is halved (equivalent to two resistors in parallel).

Calculating the resistance of a section of conductive material is simple if the shape is rectangular. Let's say you have a rectangle of 100 $\Omega/\square$ that is 1 mm wide by 20 mm long. That is equivalent to 20 1 mm × 1 mm squares in series. The resistance is 100 Ω * (20 mm / 1 mm) = 2000 Ω. What if the rectangle is not a perfect number of squares? What if it is 1 mm by 1.3 mm? We can think of this as one rectangle that is 1 mm by 1 mm connected to a second rectangle that is 1 mm by 0.3 mm. The 0.3 rectangle can be broken down into several smaller squares as shown in Fig. 7.6.

The equivalent circuit for this model is shown in Fig. 7.7, which in turn simplifies to Fig. 7.8.

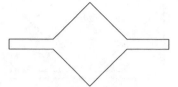

Fig. 7.9 Diamond shape

Fig. 7.10 Diamond node

The equivalent resistance is 1.3 *R*. That is the same value we get if we just divided the length by the width: $R * (1.3 \text{ mm} / 1 \text{ mm}) = 1.3 \ R$. So even with rectangular shapes that are not an exact number of squares, the resistance is $R * (L / W)$.

What about non-rectangular shapes? This is where things get difficult. The only way to determine the resistance of an arbitrarily shaped conductive sheet is with mathematical modeling. However, as we will learn later, most touch sensor electrodes are simple diamond shapes connected by straight necks (Fig. 7.9).

Can we calculate (or at least estimate) the resistance of this shape? We can, but it gets a bit tricky. Feel free to skip ahead to the answer (Eq. 7.18) if you want avoid a lot of algebra and trigonometry.

First, we need to analyze the shape of an actual node, in other words the crossing of an X and Y electrode. For reasons that will explained in the next chapter, the actual node in a diamond pattern is really two half diamonds connected by a neck (Fig. 7.10).

For the resistance analysis, we will model this as three rectangles and four triangles (Fig. 7.11).

We know how to calculate the resistance of the three rectangles. But what about the triangles? The trick is to think of the triangles as parallel resistors to the necks that run between them. The equivalent circuit would be similar to Fig. 7.12.

Once we know the resistance of the triangles, we can plug them into the equivalent circuit along with the neck resistances and calculate the total resistance. To calculate the triangle resistance, we start by breaking the triangle up into rectangles (Fig. 7.13).

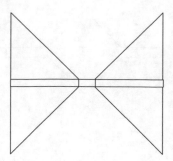

Fig. 7.11 Diamond node sections

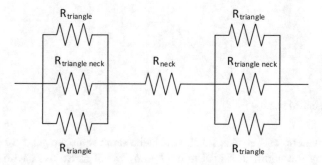

Fig. 7.12 Node equivalent circuit

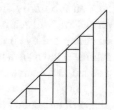

Fig. 7.13 Rectangles

Each rectangle represents a resistor. The resistance of each rectangle is the length of the rectangle divided by its height. We can calculate the height of each rectangle using a little trig. Remember that the tangent of an angle is equal to the opposite side divided by the adjacent side. If we know the angle and the length of the adjacent side, we can turn that around to calculate the height (Fig. 7.14):

$$\tan(\theta) = \frac{h}{l} \tag{7.4}$$

$$h = l \tan(\theta) \tag{7.5}$$

To calculate the height of each rectangle in our triangle, we just need to know how far that rectangle is from the corner. In this example, we have divided up our

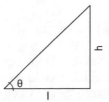

Fig. 7.14 Tangent of an angle

triangle into seven rectangles (actually six rectangles plus an empty space in the corner that is the same width as each rectangle). So each rectangle is 1/7 of the total length. The height of the first three rectangles are calculated in Eqs. 7.6, 7.7, and 7.8:

$$h_1 = \frac{1}{7} l \tan(\theta) \tag{7.6}$$

$$h_2 = \frac{2}{7} l \tan(\theta) \tag{7.7}$$

$$h_3 = \frac{3}{7} l \tan(\theta) \tag{7.8}$$

The resistance of the first three rectangles are calculated in Eqs. 7.9, 7.10, and 7.11 ($R_\blacksquare$ represents the resistance of a square of the material):

$$R_1 = \frac{l_1}{h_1} R_\blacksquare = \frac{\frac{1}{7}l}{\frac{1}{7} l \tan(\theta)} \tag{7.9}$$

$$R_2 = \frac{l_2}{h_2} R_\blacksquare = \frac{\frac{1}{7}l}{\frac{2}{7} l \tan(\theta)} \tag{7.10}$$

$$R_3 = \frac{l_3}{h_3} R_\blacksquare = \frac{\frac{1}{7}l}{\frac{3}{7} l \tan(\theta)} \tag{7.11}$$

A quick glance at the three resistance calculations shows an obvious pattern. Each resistance value is just a multiple of $\tan(\Theta)$. In other words, the resistance of the triangle only depends on the angle, not the dimensions of the triangle. A 45° triangle with sides of 1 mm will have the same resistance as a 45° triangle with sides of 10 mm. This makes some sense as it is a close analog to the rule that the resistance of a square does not depend on the size of the square. So the total resistance of the triangle is the sum of the multiples of the tangent of the angle:

$$R_T = \frac{R_\blacksquare}{\tan{(\theta)}} + \frac{R_\blacksquare}{2\tan{(\theta)}} + \frac{R_\blacksquare}{3\ \tan{(\theta)}} + \ldots \tag{7.12}$$

The number of summed terms is equal to the number of rectangles we decide to include minus 1. For seven rectangles we sum six terms. For 30 rectangles we sum 29 terms up to 1/29 tan(Θ). Of course we can factor out the resistance and tangent to get:

$$R_T = \frac{R_\blacksquare}{\tan{(\theta)}} \left(\frac{1}{1} + \frac{1}{2} + \frac{1}{3} + \ldots \right) \tag{7.13}$$

You may recognize the sum as the Harmonic series. And if you are familiar with the Harmonic series, you may already see a problem. Going back to our concept of filling our triangle with more and more rectangles to get a more accurate value, it would be logical to extend this and say we want an infinite number of rectangles. That means we want an infinite sum of the Harmonic series. The problem is that the Harmonic series is divergent. It does not have a sum – it just keeps getting bigger. But obviously the resistance is not infinite, so what is going on?

The problem is that our model of the resistance of a triangle as the sum of a set of parallel resistances is only an approximation. Therefore, the calculation that comes out of that model is only an approximation. The good news is that the approximation can still be useful. By comparing more accurate resistance calculations from mathematical modeling software to our Harmonic series approximation, we find that summing around 25 terms of the Harmonic series gets us very close to the correct resistance for a variety of angles close to 45°. The approximation is not very good at very wide angles like 80° or very narrow angles like 10°, but fortunately due to the symmetric nature of touch sensors, those angles are not common in electrode patterns. The sum of the first 25 terms of the Harmonic series is 3.816. So the resistance of a one of our triangles is approximately:

$$R_T = \frac{3.816\,R_\blacksquare}{\tan{(\theta)}} \tag{7.14}$$

When designing our touch sensor it is easier to work with the length (l) and height (h) of the triangle than the tangent of the angle. That gives us:

$$R_T = \frac{3.816\,l\,R_\blacksquare}{h} \tag{7.15}$$

The total node resistance is the resistance of the middle neck plus two times the parallel resistances of the triangle necks and triangles:

Fig. 7.15 Node dimensions

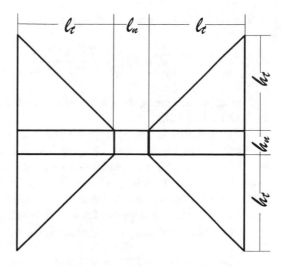

Fig. 7.16 Node example
dimensions

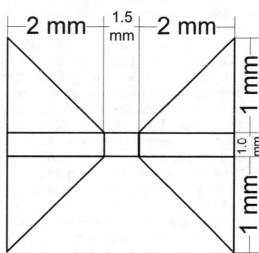

$$R = R_\blacksquare \left(R_{\text{neck}} + 2 * \left(R_T \parallel R_T \parallel R_{\text{triangle neck}} \right) \right) \tag{7.16}$$

If we plug the dimensions of Fig. 7.15 into Eq. 7.16 along with the triangle resistance from Eq. 7.15 then do some algebraic simplification, we find that the node resistance is:

$$R = R_\blacksquare \left(\frac{7.632\, l_t}{3.816\, h_n + 2\, h_t} + \frac{l_n}{h_n} \right) \tag{7.17}$$

For a concrete example, let's use the dimensions in Fig. 7.16 with a sheet resistance of 80 $\Omega/\square$.

The node resistance for this design is approximately:

$$R = 80\left(\frac{7.632*2}{3.816*1 + 2*1} + \frac{1.5}{1}\right) = 330\ \Omega \qquad (7.18)$$

If there are 25 nodes, then the total resistance of the electrode is approximately 25 * 330 = 8249 Ω.

Sheet resistance is a key metric for conductive materials used for touch sensors. An ideal conductive material for a touch sensor would have a sheet resistance of less than 1 Ω/□ with a transmissivity of greater than 95% and haze less than 1%. It would also be infinitely flexible so it could be formed into any shape. Unfortunately, there is no conductive material that meets all of those criteria. That is one reason there are multiple types of conductive materials being used in touch sensors. Several popular conductive materials are listed below starting with the most popular, ITO.

Indium tin oxide, or *ITO* as it is more commonly known, is the most popular transparent conductor used in touch sensors. ITO is a mixture of two compounds, indium oxide (In_2O_3) and tin oxide (SnO_2). Different ITO compounds have a different concentration of indium oxide to tin oxide. The two compounds are combined into a homogenous block which is then used as a sputtering target. The ITO is sputtered onto a substrate, typically glass or plastic (there are other ways of depositing ITO onto a substrate, but sputtering is the most common). During the sputtering process the substrate gets coated with a sheet of ITO. The resistance of the ITO is controlled by the particular ITO compound used to make the target, by the thickness of the sputtered layer, and by the process temperature. An ITO layer on glass is typically around 200 Å thick. An ITO layer on plastic is typically 1000 to 2000 Å. An Angstrom is 10^{-10} m or 0.1 nm.

Given a particular ITO compound and substrate, the sheet resistance of the ITO is controlled by the thickness of the deposition layer and the annealing temperature. A thicker layer has lower resistance (refer back to Eq. 7.1), but it also has worse optical properties. The transmissivity decreases and haze increases. As ITO thickens, it also takes on a noticeable yellow tint. For touch screen applications, ITO resistance is typically between 60 Ω/□ and 200 Ω/□.

ITO is the most popular transparent conductor for several reasons:

- It is inexpensive.
- It is readily available.
- It is easy to pattern into custom shapes.
- It has high transmissivity and low haze.
- It is available on plastic and glass substrates.

However, ITO also has some downsides. The biggest issue is the resistance. Larger touch panels have longer electrodes which require lower sheet resistance. 100 Ω/□ or 150 Ω/□ will work for a 7 in touch sensor, but it will not work very well for a 15.6 in touch sensor. There are some ways to lower the overall resistance as the

Fig. 7.17 Carbon nanobud. (Source: https://canatu.com/technology/)

panel gets larger, but at some point the sheet resistance simply has to decrease. And while it is possible to get ITO on glass at resistances of 30 $\Omega/\square$, the optical quality is pretty poor.

Another issue is flexibility. Even though it is possible to sputter ITO onto a plastic substrate, the ITO layer is brittle and will crack when flexed past a certain point or when flexed over and over again. It is also difficult to get low ITO resistances on plastic substrates since lowering the resistance requires high temperatures which exceed the maximum temperature of most plastic substrates. For this reason ITO is not a good choice for flexible touch sensors that are required for the new family of curved displays, particularly in the wearables market. ITO on a plastic substrate is also not a good choice for outdoor applications due to the birefringence effect of most plastics (refer to Chap. 6 for more information on birefringence).

Another potential issue with ITO is global supply. For the last 10 years there has been a general concern that ITO was going to become scarce which would increase prices. This has not actually happened, but the threat of it, along with some of the other issues with ITO, has driven a number of companies to develop alternatives to ITO. There are some dozen or so ITO alternative materials available today, each with their own pros and cons. Some of the more popular ITO alternatives are reviewed below.

Carbon nanobuds, a variation of carbon nanotubes, are cylindrically shaped chains of carbon molecules with spherical fullerene molecules essentially stuck on the side of the tubes (Fig. 7.17).

Carbon nanobuds have low electrical resistance. They can also be formed into very long chains when deposited on a substrate. The nanobuds form a random network of molecule chains on the substrate providing multiple conductive paths across the sheet (Fig. 7.18).

The chains themselves are very narrow, leaving a lot of empty space for light transmission. When light does hit one of the nanobud chains, it is absorbed. This limits undesired optical effects like reflection and haze. And since the molecule

Fig. 7.18 Carbon nanobud
network. (Source: https://
canatu.com/technology/
#nanobud)

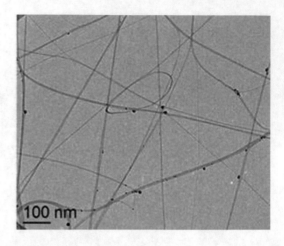

Table 7.1 Optical qualities
of carbon nanobuds

Characteristic	Typical value
Resistance	100–200 $\Omega/\square$
Transmissivity	90–98%
Haze	Less than 0.5%

Fig. 7.19 Metal mesh

100 μm to 500 μm

chains themselves are only a few nanometers wide, they are not visible to the naked
eye.

One of the biggest benefits of sheets of carbon nanobuds is their flexibility. As the
substrate bends and flexes, the nanobuds migrate over one another maintaining their
electrical network. This makes carbon nanobuds especially suited to applications
requiring a curved or constantly flexing touch sensor. Carbon nanobuds can also be
molded into a variety of shapes, enabling touch on nontraditional surfaces. Refer to
Table 7.1 for a list of the properties of carbon nanobuds.

Fig. 7.20 Moiré pattern

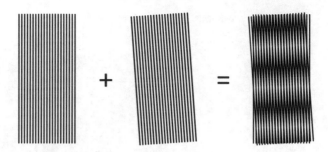

Fig. 7.21 Wire mesh and pixel mask

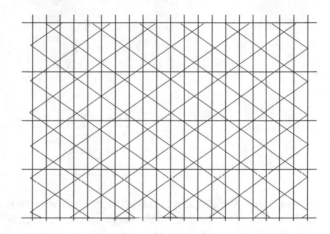

Table 7.2 Optical qualities of metal mesh

Characteristic	Typical value
Resistance	1–10 Ω/□
Transmissivity	90% or better
Haze	Less than 1%

Metal mesh, also known as *wire mesh*, is a mesh of wires connected in a diamond pattern and placed onto a substrate. The mesh looks like a shrunken window screen (Fig. 7.19).

Because the mesh is made of metallic wires (silver and copper are the most common choices), the resistance is very low, typically less than 10 Ω/□. This makes metal mesh ideal for large touch sensors.

One of the issues with metal mesh is that a metal mesh touch sensor in front of a TFT display can often create a noticeable Moiré pattern. A Moiré pattern is an interference pattern created by two overlapping, transparent patterns (Fig. 7.20).

In this case, the two patterns are the diamond pattern of the metal mesh and the pixel mask of the TFT (Fig. 7.21).

Moiré effects can be eliminated by adjusting the metal mesh dimensions so that they line up with the pixel mask. However, this means that the metal mesh material has to be customized to match the particular TFT it is being paired with. For this

Fig. 7.22 Silver nanowire

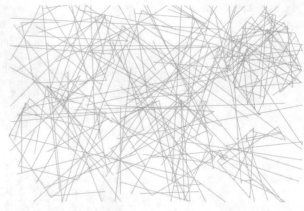

1 µm to 5 µm

Table 7.3 Optical qualities of silver nanowire

Characteristic	Typical value
Resistance	10–100 Ω/□
Transmissivity	95% or better
Haze	Less than 2%

reason, metal mesh is most often used in large volume touch sensor applications. Refer to Table 7.2 for a list of the properties of metal mesh.

Silver Nanowire is similar to metal mesh in that it uses metal wires. But unlike metal mesh which has a regular pattern, nanowire material is a random scattering of wire sections that looks like the children's game Pick-Up Sticks (Fig. 7.22).

The resistance of the material depends on the density of the wires. Low density materials have resistances in the 50 Ω/□ range, while high densities have resistances less than 10 Ω/□. Of course, higher density materials also have less transmittance and more haze. But even at relatively low resistances silver nanowire has higher transmissivity than ITO, typically greater than 95%. And since the wires are randomly scattered, they do not create Moiré effects like metal mesh. Refer to Table 7.3 for a list of the properties of silver nanowire.

There are several other types of transparent conductors currently available including graphene, transparent polymers (PEDOT:PSS), and silver nanodrops. There are also several startups working on other types of transparent conductors that are still in the development phase. A lot of these materials have limited advantages over ITO and were developed only because market forecasters have been claiming for years that ITO prices were going to increase dramatically as global supply failed to keep up with demand. That has not happened. But that does not mean it will not eventually happen, and if (or when) ITO prices do increase, touch sensor manufacturers will begin looking at these various ITO replacement materials with much more interest.

Chapter 8
Electrode Patterns

Now that we have a good understanding of the basic principles of a touch sensor and a list of transparent conductors we can use to form our capacitors, imagine for a moment how we might actually build this sensor. The simplest way would be to use two layers of transparent conductors. The top layer would have the top capacitor plates and the bottom layer would have the bottom capacitor plates (Fig. 8.1).

This would be an exact physical translation of the concept of a grid of capacitors. But there are several issues with it. First, the capacitors would be comprised of two layers of conductive material stacked on top of each other. This reduces the optical clarity twice as much as a single layer of conductor. The area surrounding each capacitor would be clear of any conductive material. This difference in optical quality between each capacitor and the area around it would make the electrode patterns more visible.

Second, most of the electric field would be captured between the plates instead of extending up to the top surface of the sensor. Figure 8.2 shows the electric field lines of a standard plate capacitor.

Some of the field extends above the top plate and some below the bottom plate, but the height of the field outside of the two plates is limited. There is also some electric field extending from the edges of the two plates. This is called a fringe field. When two plate capacitors are next to each other, their fringe fields meet in between the two capacitors (Fig. 8.3).

We can see in Fig. 8.3 that the fringe field between the plates does not extend very far past the height of the top plate. Also the two competing fields create a nonuniform electric field between the capacitors.

We can solve the optical issue and the electric field issue by spreading out the capacitor plates so that they are not directly on top of each other. For example, we could simply shift each row by one half of the pitch of the rows and shift each column by one half pitch of the columns. That gives us a checkerboard pattern like Fig. 8.4.

The electric field for this side-by-side arrangement is shown in Fig. 8.5.

© Springer International Publishing AG, part of Springer Nature 2019

T. Gray, *Projected Capacitive Touch*, https://doi.org/10.1007/978-3-319-98392-9_8

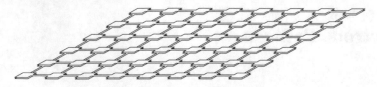

Fig. 8.1 Plate capacitor electrode pattern

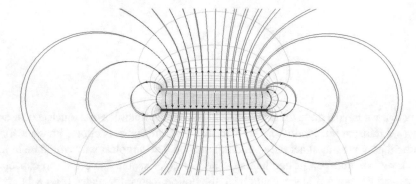

Fig. 8.2 Plate capacitor electric field

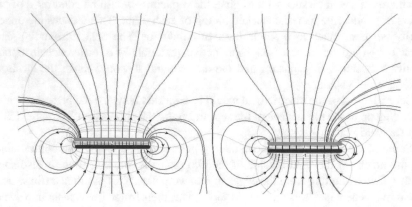

Fig. 8.3 Electric field of side-by-side plate capacitors

The electric fields are no longer captured between the plates. Instead the fields extend further into space. They are also more tightly coupled from plate to plate.

Going back to the electrode pattern show in Fig. 8.6, we still have large areas of blank space between the capacitor plates. The contrast between these blank areas and plates could still be visible. To solve this problem, we really need the plates to fill up as much space as possible. We can accomplish this by rotating the plates 45° and making them large enough that the plates are very close together. This creates the common diamond pattern used in most projected capacitive touch sensors (Fig. 8.6).

We do not want the plates to overlap, so we need some gap between them. The size of that gap affects the height of the electric field between the plates. That can be

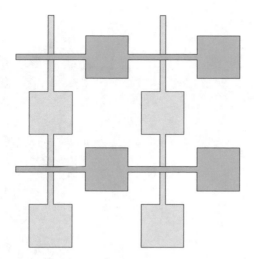

Fig. 8.4 Top and bottom plates moved

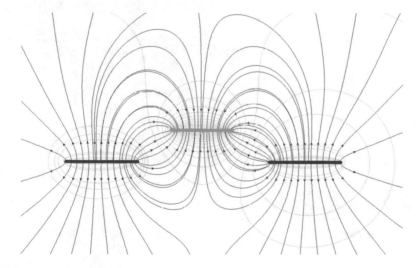

Fig. 8.5 Spread plates electric field

useful. If we need to use a thick cover lens, or if the user needs support for thick gloves, we can widen the gap and increase the size of the electric field. On the other hand, the smaller the gap, the harder it is to discern the pattern. So there is a balancing act between using the smallest gap possible to minimize the visibility of the pattern and using the widest gap possible to create a large electric field.

One common misconception about this diamond design is what actually constitutes the capacitor at each node (i.e., each X and Y electrode crossing). Figure 8.7 shows a larger diamond pattern with red boxes indicating the size of each individual node and red dots indicating the centers of the nodes.

Fig. 8.6 Diamond pattern

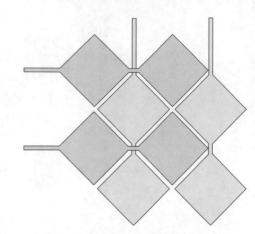

Fig. 8.7 Physical node location

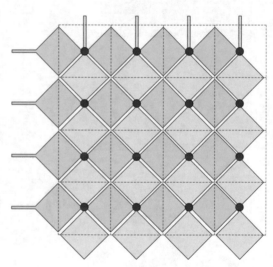

Fig. 8.8 Node structure

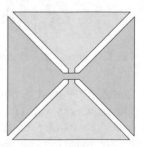

Looking at the structures in each box, it is clear that a single node looks like Fig. 8.8.

In other words, each diamond should really be thought of as two connected triangles. In a horizontal electrode the left triangle is part of the left node, and the

Fig. 8.9 Full diamond
pattern

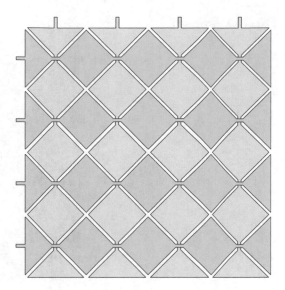

right triangle is part of the right node. The same applies to top and bottom triangles
for vertical electrodes. Thinking of the diamonds as two triangles also gives us some
guidance on how to deal with the ends of the electrodes where we previously had
some protruding diamonds and some blank spaces. We just add or remove triangles
to the ends of the electrodes so that each node has the same structure (Fig. 8.9).

In this pattern, the actual plate capacitor is the tiny area where the horizontal neck
passes over the vertical neck of each node. We could use the standard formula from
Chap. 2 to calculate the value of that capacitor (Eq. 8.1):

$$C = \varepsilon \frac{A}{d} \tag{8.1}$$

We just need to know the area of crossover between the necks (A), the distance
between the layers (d), and the dielectric of the material between the layers (ε). If we
were to calculate this capacitance and compare it to the empirically measured
capacitance of a node, we would find a significant difference between the two.
The reason for the difference is that the fringe field between the edges of the triangles
has a large influence on the capacitance. Looking back at the electric field in Fig. 8.6,
it is clear that a large amount of field is present between the edges of the plates. There
is no easy formula for calculating capacitance that includes a large fringe field
component. That is another reason that mathematical modeling is often required to
determine the capacitance of a node and how much that capacitance is going to
change when a finger is present.

The diamond pattern is the most common electrode pattern used in touch sensors.
However, it is not the only pattern. The second most common pattern is Manhattan
or Flooded X. In this pattern the drive electrodes, which are always on the bottom

Fig. 8.10 Manhattan

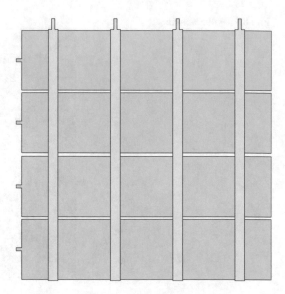

layer, are thick rectangles, while the sense electrodes, which are always on the top, are skinny rectangles (Fig. 8.10).

The biggest benefit of Manhattan is that the wide drive electrodes on the bottom layer shield the sense electrodes from radiated noise emanating from the display. That is why it is also called Flooded X because the X layer is essentially a flooded ground plane.

There are some problems with the Manhattan pattern. The first is that most of the electric field comes from the plate capacitance where each skinny sense electrode passes over a drive electrode. This places tight constraints on the thicknesses and materials used to construct the touch sensor. This is due to the limited size of the electric field directly above the electrodes. The key parameter is the ratio of the distance between the finger and the drive electrode to the distance between the drive and sense electrodes. If the ratio is too small, then the touch sensor is not sensitive enough and will not work properly. If the ratio is too large, the sensor is too sensitive and will have a poor signal-to-noise ratio and may even have false touches from noise. This causes problems not just for the initial design but for any changes. For example, let's assume that we have a well-designed Manhattan sensor with good sensitivity. We want to take that sensor and add a cover lens to it. This increases the distance between the drive electrode and the finger. To keep the panel in balance, we also have to increase the distance between the drive and sense electrodes. This may be difficult or impossible depending on the construction of the panel and the thicknesses of the various adhesives and conductive substrates available.

Another issue with Manhattan is that the touch sensitivity falls off very quickly between sense electrodes. If the panel is not designed well, a small finger may not register in between sense electrodes. There are some ways to mitigate this by adding extensions to the sense electrodes, but this can only help so much (Fig. 8.11).

Fig. 8.11 Manhattan with cross bars

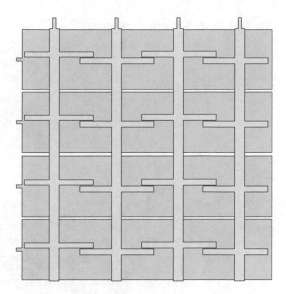

The final issue with Manhattan is that it only supports mutual capacitance. As mentioned in a previous chapter, mutual capacitance is the most commonly used method for capacitive touch sensors, so this may not seem like much of a determinant. But you may also remember that some touch controllers combine mutual and self measurements to provide greater sensitivity for features like glove and water detection. These features cannot be used with a standard Manhattan pattern. When using self capacitance the electrodes all have to be the same size and shape (meaning the same impedance). Since all the electrodes are driven simultaneously, any difference in impedance from one electrode to the next will result in huge variations in the electric field making it difficult to pinpoint the location of a touch.

So far we have only discussed sensor patterns that exist on two separate conductive layers. Is it possible to make a mutual capacitance sensor using a single conductive layer? The answer is yes! There are two main methods for building touch sensors on a single conductive layer. The first is to use the standard diamond pattern in ITO with an additional feature called micro-crossovers. These are essentially tiny conductive bridges that act as overpasses for one of the electrode layers. These micro-crossovers allow the necks on one layer to pass over the necks on the other layer. Figure 8.12 shows a top down view and a side view of what this looks like for a single node.

There are three major problems with micro-crossovers. First, the micro-crossover is a solid metal conductor (with insulator underneath) and therefore blocks light. So it reduces the transmissivity and increases haze. Second, the equipment to place the micro-crossovers is very expensive and is only economical for large volume touch sensors. Third, micro-crossovers are typically used on PET sensors. This can be a positive attribute as PET sensors are thinner and lighter than glass, but it is also a detriment as glass has better transmissivity, lower haze, lower ITO resistance, and is not birefringent.

Fig. 8.12 Micro-crossover

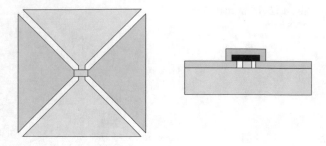

Fig. 8.13 True
single layer
design

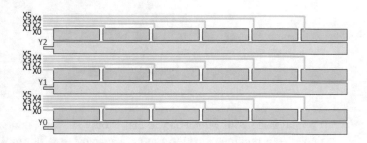

The second way to produce a single layer touch sensor is to use what is called a true single layer pattern. This is a pattern that does not require any electrodes to cross over each other. There are a variety of true single layer designs, most of them proprietary to specific companies. But most of them follow the same basic principle. The sense (Y) electrodes run in straight lines along the panel. The drive (X) electrodes are routed in channels between the sense lines. The tricky thing is that *all* of the drive lines have to be run down each channel (Fig. 8.13).

There are several difficulties with this pattern. First, the number of drive lines is limited by the maximum distance between the sense lines. This places a maximum limit on the number of drive electrodes which in turn limits the overall size of the touch sensor. Most true single layer designs are limited to 5 in diagonal displays (some proprietary designs can go up to 7 in diagonal). The second issue is the number of connections needed. On a standard sensor, the number of necessary connections is X + Y (not including a few grounds and maybe a few shield lines). For a true single layer design, the number of connections is X * Y. We will use a standard 4.3 in diagonal sensor as an example. The active area is 95 mm × 53 mm. We will use a 7 mm pitch for the electrodes, which means we need 14 × 8 electrodes. For a standard sensor, we would need 14 + 8 = 22 electrode connections. For a true single layer design, we would need 14 * 8 = 112 electrode connections! This complicates the design and manufacturing and increases the overall cost. Another issue is that the electric fields vary significantly from node to node. These variations can be smoothed out to some extent by altering the electrode shapes. This requires

Fig. 8.14 Diamond vertical
layer

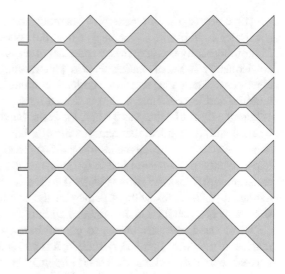

Fig. 8.15 Diamond
horizontal layer

sophisticated electrostatic modeling which is why these patterns are often proprietary. For these reasons, true single layer designs are not used very often.

Figure 8.9 showed a two-layer diamond pattern. In Figs. 8.14 and 8.15 the pattern
is split into the individual patterns for each of the two layers.

How are these patterns formed from the conductive materials mentioned in the
previous chapter? To some extent that depends on the material. Some ITO alternatives can be printed directly into the desired patterns. But what about sheet-coated
conductors like ITO?

There are two major patterning methods for creating the electrode shapes on substrates coated with a conductive material: photolithography and laser etching. Photolithography is a very messy chemical process requiring a lot of equipment (and consequently a lot of space). First a pattern mask is made in the shape of the electrodes. Then the sheet of conductive material is coated with a photoresist material and baked. Next a light is shined through the mask onto the photo resist. Where the light hits the photo resist, the resist becomes hardened. The sheet is then washed with a chemical that removes the non-hardened resist. Next it is placed in a chemical etch bath which etches away all of the exposed conductive material. This leaves conductive material just in the shape of the electrodes (with hardened photo resist on top). Next the sheet is washed with a second chemical which removes the hardened photo resist without damaging the conductive material underneath. The sheet now has conductive material etched into the desired electrode shapes.

There are a few advantages to photolithography. First, the size of individual features can be very small. For example, it is possible to etch a line as small as 30 μm between conductive elements. This is important because smaller etch lines are harder to see. The second benefit is that *all* of the material between the electrodes is removed. This is helpful because leaving floating conductive material between the electrodes can affect the capacitive measurements. But as mentioned before, photo-lithography is complicated and dangerous and requires a lot of equipment and a lot of space. It also cannot be used on some of the ITO alternative materials mentioned in the previous chapter. That is why laser etching has become the preferred method for electrode patterning.

With laser etching, the sheet of conductive material is placed inside a large chamber. A laser is then used to essentially blast the conductive material off of the sheet. This is a much simpler process than photolithography, requires a lot less equipment, and takes up a lot less space. One downside of laser etching is time. The laser has to trace around every electrode shape. Therefore, the more electrodes there

Fig. 8.16 Laser etched
diamond pattern

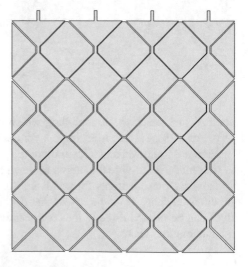

are and the larger the touch panel, the longer the process time. It also means that the material between electrodes is not removed as blasting away all that material would take an enormous amount of time. Instead, the laser etches lines through the remaining material to break it up into individual islands. This helps reduce the capacitive coupling between the floating conductive material and the electrode. But it also means that almost the entire top and bottom layers are coated with conductor, so the light has to pass through two layers of conductive material instead of one. After laser etching, the pattern of the conductive material looks something like Fig. 8.16.

In between the electrodes are essentially floating diamonds of material. Each diamond is separated from its neighbor with a small etch line. As with photolithography, one benefit of laser etching is that the etch lines are fairly small, around 50 µm. With some other patterning technologies, like silk screening, the etch lines can be as large as 200 µm.

In addition to creating the electrodes, we also need some way to route the electrode connections to the edge of the sensor. We could create these traces in ITO, but since these traces are not directly over the display they do not have to be transparent. So instead of using ITO we can use a much lower resistance material like silver. These perimeter traces are discussed in the next chapter.

Chapter 9
Perimeter Traces and Tails

So far we have only talked about the electrodes that make up the actual touch sensor. These electrodes need to have some way of connecting to the outside world in order to be useful. These outside connections are provided via perimeter traces and flexible printed circuit (FPC) tails. To begin our discussion, let's first consider the physical outline of a specific LCD. We will use a 7 in diagonal display for our example which has an active area of approximately 150 mm × 90 mm. What we mean by "active area" is the area defined by the outermost pixels on the display. There will be a small black gap around that area, usually about 1 mm. The area including the black gap, which would be 152 mm × 92 mm, is known as the viewable area. Outside of the viewable area is a surrounding bezel which is most commonly made of tin (Fig. 9.1).

Our touch sensor is going to sit on top of this display. In order for our touch coordinates to match our LCD coordinates, we need the active area of the touch sensor to align with the active area of the LCD. In other words, we want our electrodes to be 150 mm × 90 mm (Fig. 9.2).

We need to route our electrodes to the edge of the touch sensor. Ideally we would like this routing to be done with a low resistivity conductor, something like silver. But we also do not want any visible traces in the viewing area. So at the ends of our electrodes we will add a small stub of the transparent conductor material to get us outside the viewable area of the LCD. Then we will make a large pad that we can bond to (Fig. 9.3).

Now that we are outside the viewable area, we can switch to an opaque conductor like silver and run our traces to the electrode pads. We could continue to use just ITO, but the resistance would increase quite a bit since we are using long, skinny traces around the perimeter. Since these traces are not visible, we can reduce the resistance by using an opaque conductor with a very low resistance like silver. These opaque, metalized traces are called perimeter traces. The amount of space required for the perimeter traces depends entirely on the process used for putting down and patterning the conductive material. One common technique is to silkscreen the entire area with the conductive material, and then etch out the individual traces with a laser (Fig. 9.4).

© Springer International Publishing AG, part of Springer Nature 2019
T. Gray, *Projected Capacitive Touch*, https://doi.org/10.1007/978-3-319-98392-9_9

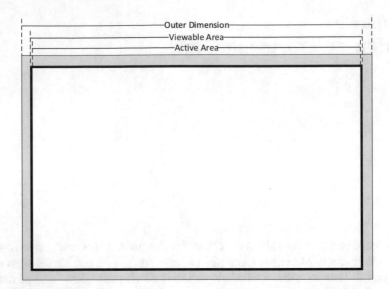

Fig. 9.1 LCD viewable and active areas

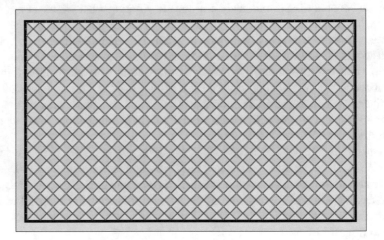

Fig. 9.2 LCD with electrodes on top

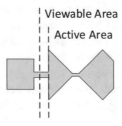

Fig. 9.3 Electrode pad

Fig. 9.4 Silver trace
etching

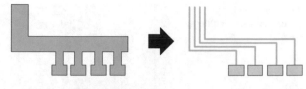

Fig. 9.5 Perimeter traces

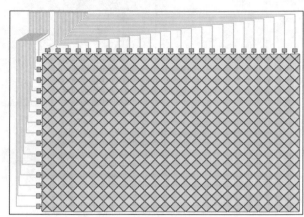

This results in tight pitches for the traces (50–100 μm). It also has the benefit that, if the laser is also being used to pattern the transparent conductor, then both etch patterns can be executed in the same step. Another option is to silkscreen the individual traces, but this results in much larger trace pitches (150–250 μm) and thus takes more space.

The perimeter traces are run to another set of bonding pads on the edge of the touch sensor. In theory, the two sets of perimeter traces, one for the horizontal electrodes and one for the vertical electrodes, can be run to separate edges. In practice, they are usually right next to each other (Fig. 9.5).

In addition to the electrode traces, there are often ground traces run around the perimeter. These are typically for ESD protection, but they also help with creating a balanced capacitive load on the outside electrodes. They are wider than the trace electrodes and are often run underneath the bond pads of the opposite layer. In a two layer sensor these ground electrodes are run on both layers. Figure 9.6 shows the electrode pattern and traces for a single layer including the ground trace (keep in mind that the electrodes are shown in yellow in the figure but are actually transparent).

All the traces terminate on the edge of the sensor in another set of bond pads. These pads are fairly narrow and long (Fig. 9.7).

A flexible printed circuit (FPC) tail is adhered to these pads. The tail is a polyimide substrate coated with copper traces. The end of the tail has exposed pads the same size and shape as the bond pads on the edge of the touch sensor. The other end of the tail also has exposed pads designed to be inserted into a zero insertion force (ZIF) socket or bonded directly to a PCB (Fig. 9.8).

Fig. 9.6 One layer
perimeter traces including
ground

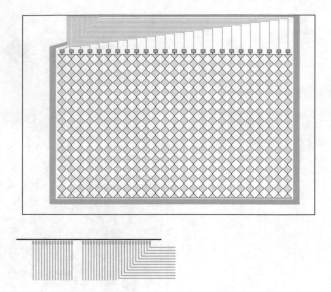

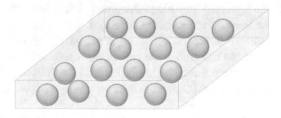

Fig. 9.7 Bond pads

Fig. 9.8 Polyimide tail

Fig. 9.9 ACF

The tail is bonded to the touch sensor using a special bonding adhesive called anisotropic conductive adhesive or ACA. It is also often referred to as ACF for anisotropic conductive film. ACA contains small conductive gold balls (3–10 µm diameter) suspended in a resin (Fig. 9.9).

Fig. 9.10 Bonded FPC
(not to scale)

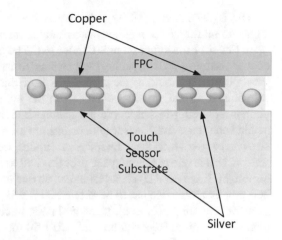

Fig. 9.11 Full
sensor

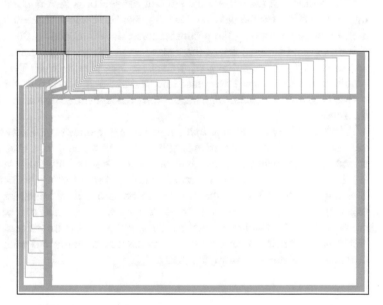

The ACF resin is dispensed on top of the silver bond pads on the edge of the touch sensor substrate. The tail is then placed on top of the resin with the traces facing toward the resin. A hot metal bar is then lowered onto the tail and pressure is applied. After a few seconds the resin sets. Balls that lie between the silver on the touch sensor and the pad on the tail are compressed and form a conductive path between the two metals. The balls are distributed within the resin in such a way that they do not touch each other. This prevents side-to-side shorts (Fig. 9.10).

Figure 9.11 shows the final two-layer sensor with bonded tails (the electrodes are not shown in this figure so you can see the visible components of the sensor).

   This is just a theoretical example. In practice, the traces are often routed at various angles to minimize the required space and the number of right angles.

   In Fig. 9.11 the leftmost tail is bonded to the bottom layer of the sensor and has its copper facing up. The rightmost tail is bonded to the top layer and has its copper facing down. It should be clear from Fig. 9.11 that the overall size of the sensor is constrained by the space required to run the perimeter traces. If tighter electrode pitch is required, which is helpful for features like fine tip stylus, then more traces are needed and the overall sensor dimensions must be larger. If a larger electrode pitch size can be used, then there are fewer electrodes, less perimeter traces, and a smaller sensor. This is one reason why the process used to create the perimeter traces is very important. If we need 20 electrode traces on one edge and the required trace pitch is 200 µm, then the edge has to be at least 20 * 200 µm = 4 mm wide (plus room for grounds, bonding pads, etc.). If the perimeter trace pitch can be reduced to 50 µm, then the minimum edge width is only 20 * 50 µm = 1 mm.

   The position of the tails on the edge of the sensor is also important. There is an optimal position for the tails, somewhere near the edge of the sensor, at which the area required for routing the perimeter traces is minimized. If the tails have to be shifted closer to the center of the sensor, then the perimeter trace area has to be increased and the overall size of the sensor increases as well. Typically, the tails would not be shifted past the center line. If they needed to be that far over, it would be easier to route the outer edge traces along the right edge instead of the left edge of the senor.

   The tails are typically kept to a single layer to reduce cost. However, multilayer tails can be used to meet certain requirements. For example, a shield layer can be added on either side of the tail (two layers total) or to both sides of the tail (three layers total). This helps to block radiated noise from infecting the electrode signals. The tail may have vias so that traces can be rerouted or joined together. This is especially helpful on true single layer designs where the tail can be used to cross connect all the X0 electrodes, all the X1 electrodes, and so on (refer to the chapter on electrode patterns). Some two layer sensors use a single tail with a Y split for bonding to the touch sensor (Fig. 9.12).

Fig. 9.12 Y tail

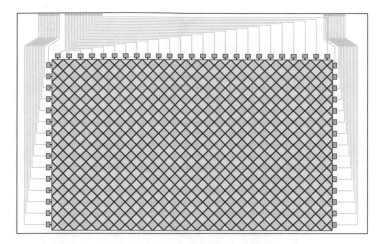

Fig. 9.13 Double routing

The Y tail can be single layer in which case the two conductive sensor layers must be facing the same direction, or the tail can be two layers with one end of the tail facing down and the other Y end facing up (of course this would require vias to move one set of traces from the bottom copper layer to the top or vice versa). The sizes and shapes of the tails are one of the most common items requiring customization as the tails have to route around mechanical constraints, through cutouts, over other system components, be lengthened to reposition the controller board, etc.

Some larger designs require what is called double routing on one or both axes to reduce the overall electrode resistance. Double routing means that the perimeter traces are routed to both ends of the electrodes (Fig. 9.13).

Of course, this complicates the tail design. In Fig. 9.13, the extra perimeter traces are shown routed to a third tail on the right side of the sensor. That also complicates the controller board design. A better option is to route the new traces around the bottom of the sensor and along the left edge so that they are next to their companion traces. But this requires the perimeter area to be expanded along the bottom and left edges, expanding the size of the entire sensor. Another option, especially useful when double routing is required on all four edges, is to have tails come off all four edges and meet in the center behind the LCD (Fig. 9.14).

Sometimes called the "Christmas Present Design" because all the traces meet near the center of the LCD where they are inserted into the controller board, this option reduces the perimeter area needed for trace routing (Fig. 9.14 shows straight traces for clarity; in practice, the traces are run at various angles to minimize the perimeter edge). Figure 9.15 shows a real-life example of a double routed design that had several mechanical constraints requiring the tails to be split into multiple tails.

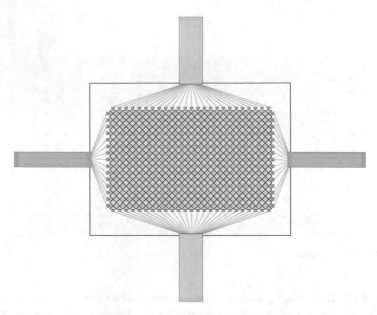

Fig. 9.14 Tails on all four edges

Fig. 9.15 Complicated multi-tail routing

The key to perimeter trace routing and tail placement is to minimize the area needed for routing the perimeter traces as much as possible. The constraints are:

- The minimum trace width and pitch which is based on the perimeter trace process
- Double routing (if needed)
- Routing constraints on the traces themselves (e.g., do not run an X trace parallel to a Y trace)

- Required bonding pads sizes
- Mechanical constraints that restrict the tail locations
- Mechanical constraints on the overall size of the sensor
- The size and placement of the touch controller PCB

An experienced touch designer can work within these constraints to minimize the perimeter area needed around the sensor and minimize the overall sensor dimensions.

We now have a good understanding of the different layers that are required for a touch sensor and how the sensor traces are routed off of the sensor substrate onto FPC tails. The next chapter discusses the overall stack up of materials that comprise the display, the touch sensor, the top cover lens (if required), and the adhesives used to bond the layers together.

Chapter 10
Sensor Stack Ups

There are more than a dozen different material stack up options for a projected capacitive touch sensor. There are three main criteria that define the stack up:

1. Is the touch sensor built into the LCD or on top of it?
2. How many layers is the touch sensor?
3. Is there a cover lens?

Some LCD manufacturers actually build the components of a touch sensor directly into the LCD during manufacturing. This is called in-cell because the touch sensor electrodes are literally in the cells (or pixels) of the display. There are several different types of in-cell touch including light-based, pressure-based, and charge-based systems. Since the primary focus of this book is projected capacitive touch, we will focus on the version of in-cell that uses projected capacitive (PCAP).

A mutual capacitance in-cell touch sensor has two layers. The drive electrodes are on the same layer as the transistors that control the liquid crystals in the display. The sense electrodes are on top of the color filter glass but underneath the top polarizer (Fig. 10.1).

The benefits to this design are a thinner module stack up with improved optical quality. But there are also some difficulties. The biggest problem is electrical noise from the LCD transistors and VCOM circuit interfering with the touch sensor measurements. In this type of design, the change in capacitance that we are trying to measure is close to 1 pF. That can be difficult when the drive electrodes are adjacent to the VCOM circuits and the sense electrodes are less than a millimeter from the drive circuits. These issues can be addressed by coordinating the LCD driving with the touch sensing. This could be accomplished by having one controller drive the LCD and manage the touch signals, or by having the touch controller synchronize its scan cycles to the LCD driver. Either way requires some complicated design and tuning. And modifying the LCD mask layers to incorporate the touch electrodes can cost several million dollars. For these reasons, in-cell tends to be used only on very high volume products like cell phones.

© Springer International Publishing AG, part of Springer Nature 2019
T. Gray, *Projected Capacitive Touch*, https://doi.org/10.1007/978-3-319-98392-9_10

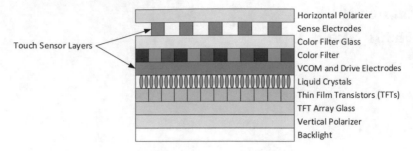

Fig. 10.1   In-cell touch sensor

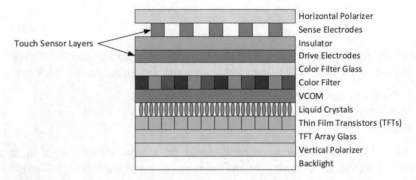

Fig. 10.2   On-cell touch sensor

There is another option for embedding the touch sensor in the LCD while reducing some of the manufacturing and tuning difficulties. In this type of design, called on-cell, the touch sensor is two separate layers separated by an insulator. These two layers are placed on top of the color filter glass below the top polarizer (Fig. 10.2).

This is a bit thicker than in-cell with slightly less transmittance, but it has the advantage that the touch sensor elements are further from the LCD electronics and are less affected by noise. Careful design and tuning are still required, but the touch sensor may not have to be synchronized to the LCD driver (although it definitely helps). And the touch sensor layers can be added without having to modify the LCD mask, a substantial cost savings. Still, on-cell requires some significant design and manufacturing investment which is why it is mainly used in mid- to high-volume consumer applications.

While in-cell and on-cell offer some optical and stack up advantages, their biggest benefit is also their biggest drawback. Because an in-cell or on-cell sensor is designed into the LCD, the touch sensor cannot be modified without significant cost. If an LCD vendor happens to have an LCD with an in-cell or on-cell touch sensor that meets your requirements, if your operating system can support the touch driver that the LCD vendor is using, if you can design your product around the mechanical dimensions of the LCD and touch sensor, and if you are not concerned about the short lifespan of an LCD and touch sensor targeted at the consumer market

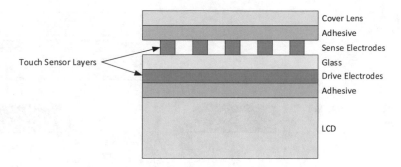

Fig. 10.3 Out-cell touch sensor

where product lifetimes are often measured in months instead of years, then in-cell or on-cell may a good fit for your application. But if you have special functional requirements like detecting touch with thick gloves or water immunity, or if you need a driver for a nonstandard operating system like Vx Works, or your enclosure has some mechanical constraints that cannot accommodate the available in-cell and on-cell modules, or if you need a five-year life span, then an in-cell or on-cell solution probably will not work for you. That is why the most popular stack up option is out-cell.

An out-cell touch sensor is essentially any touch sensor that is added onto the top of the LCD (Fig. 10.3).

Typically an out-cell touch sensor is manufactured by a company specializing in touch. The touch sensor manufacturer may purchase the LCD, integrate the touch sensor and cover lens, and provide the entire module. As another option, an original equipment manufacturer (OEM) might purchase the touch sensor and LCD from separate vendors and integrate them as they are building the product. The key point is that the touch sensor is manufactured separately from the LCD.

The biggest benefit to an out-cell design is that we can purchase whatever LCD we prefer for our application, then purchase whatever touch sensor meets our touch requirements and combine the two. We can combine a standard LCD with a touch sensor that supports water immunity and stylus. Or we can combine a high bright backlight LCD with a standard touch sensor. We can even have a custom touch sensor designed to work with a standard off-the-shelf LCD. The customizations might include placing the touch sensor tails in specific locations to get around certain mechanical features, reducing the electrode pitch to work with a small stylus, or adding privacy and anti-smudge films. And compared to an in-cell or on-cell touch sensor, a custom stand-alone touch sensor is much less expensive to design and manufacture in low volumes.

There are several possible stack ups for an out-cell sensor. An out-cell sensor can be single layer or multilayer; it can have plastic or glass substrates; there are a variety of conductive materials to choose from; a cover lens may or may not be needed; there may be additional films or coatings; etc. Several different stack up options are discussed in the following sections. The acronyms used to identify each stack up are fairly standard, although there are some occasional variations used by different

Fig. 10.4 Glass-glass
(GG) sensor

Fig. 10.5 Rear mount of
GG sensor

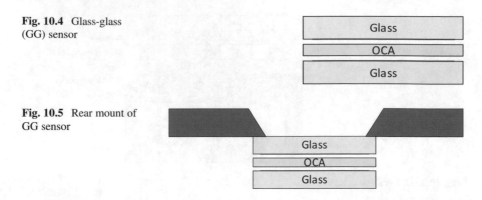

manufacturers. The stack up figures are roughly to scale vertically, meaning that the relative thicknesses of the layers is accurate. The conductive layers are shown as thick red lines because the substrate (e.g., glass) is often several thousand times thicker than the conductive material (e.g., ITO). The small gaps between the layers are for illustration only so you can easily see which layer the conductive material is on. Also, the following list does not include true single layer patterns, meaning interleaved patterns on a single conductive layer without micro-crossovers.

Glass-Glass (GG)

A glass-glass sensor is two layers of glass, each coated with conductive material (almost always ITO). The two pieces of glass are adhered to each other with a layer of optically clear adhesive (OCA). See Fig. 10.4.

While it is possible for the top glass layer to also be silkscreened, it is not common as combining silkscreened ink and patterned conductive material onto a single piece of glass is a difficult and expensive process. Consequently, a GG construction is typically used when the sensor is rear mounted in the bezel so that no decoration is required (Fig. 10.5).

Glass-Glass-Glass (GGG)

A glass-glass-glass sensor is a GG sensor with a top cover glass. See Fig. 10.6.

The top glass is a decorated cover lens. This type of sensor is typically front mounted into a shelf in the bezel, allowing for a seamless looking design similar to most cell phones and tablets. The cover glass extends past the edges of the touch sensor glass. A gasket adhesive is used to adhere the cover glass to a shelf in the bezel (Fig. 10.7).

Fig. 10.6 Glass-glass-glass
(GGG) sensor

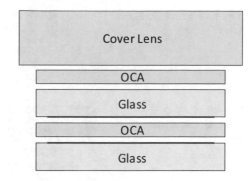

Fig. 10.7 Front-mounted
GGG sensor

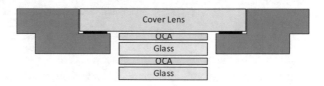

Fig. 10.8 Double-sided
glass (GG2) sensor

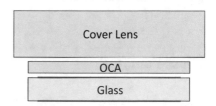

Double-Sided Glass (GG2)

A double-sided glass sensor is two pieces of glass (this type of sensor is also referred to as double-sided ITO or DITO). The top glass is the cover lens which may or may not have decorations. The inner glass layer has conductive material on both sides (again, typically ITO). The two pieces of glass are adhered with a layer of OCA. See Fig. 10.8.

A GG2 sensor is usually front mounted, but it can be rear mounted as well.

Glass-Film-Film (GFF)

A glass-film-film sensor is a cover lens with two film layers, each coated with conductive material. See Fig. 10.9.

The key benefits of a GFF sensor are that it is very thin and very light. For those reasons, GFF sensors are often used in handheld devices. Depending on what type of

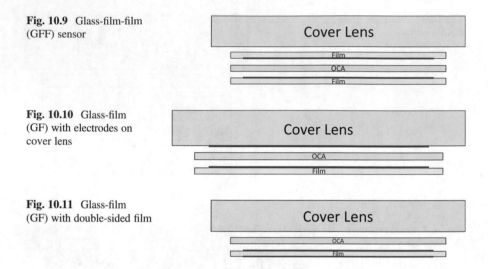

Fig. 10.9 Glass-film-film (GFF) sensor

Fig. 10.10 Glass-film (GF) with electrodes on cover lens

Fig. 10.11 Glass-film (GF) with double-sided film

film substrate and conductive materials are used, the drawbacks of a GFF sensor are degraded optical quality and birefringence. A lot of ITO alternative materials use this type of sensor construction.

Glass-Film (GF)

There are several types of glass-film structures. The first is a glass-film structure where the top electrodes are on the cover lens and the bottom electrodes are on the film. See Fig. 10.10.

As mentioned earlier, having decorative ink and a conductive layer on the back of the cover lens is a difficult, expensive process, so this type of stack up is most often used in high volume applications.

The second type of GF stack up is a double-sided film layer. See Fig. 10.11.

The third type of GF sensor is a single conductive layer with micro-crossovers (this type of construction is sometimes called SITO or single-sided ITO). See Fig. 10.12.

As with any sensor stack up that includes films, these GF constructions are thin and light but also have degraded optical proprieties compared to glass and may exhibit birefringence depending on the substrate materials.

Glass film sensors are almost always front mounted.

Fig. 10.12 Glass-film
(GF) with micro-crossovers

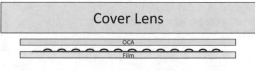
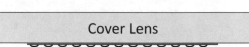

Fig. 10.13 One glass
solution (OGS)

Fig. 10.14 LCD bonded
with gasket adhesive

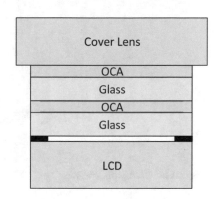

One Glass Solution (OGS)

A one glass solution is a cover lens with a conductive layer with micro-crossovers. See Fig. 10.13.

An OGS is the thinnest and optically best type of out-cell construction, but it is also one of the least common because it requires applying and patterning a conductive layer on the back of a decorated cover lens, and it requires placement of micro-crossovers. For these reasons, OGS has only been used in a handful of high volume consumer applications.

While the LCD was not shown in the stack ups, it is typically part of the touch sensor stack, meaning it is adhered to the touch sensor. The LCD may have a perimeter gasket that bonds to the touch sensor (Fig. 10.14).

This is the least expensive option, but also the worst optically as it leaves an air gap between the LCD and the touch sensor stack. The index of refraction between air and glass is very high, so the air gap causes optical distortions. A better option is to use an optical adhesive between the touch sensor and the LCD (Fig. 10.15).

There are two types of OCA used in bonding LCDs, dry and wet. Dry OCA is a sheet of adhesive material (this is the same OCA that is used to bond the touch sensor layers to each other and to bond the touch sensor to the cover lens). The OCA material may have a plastic substrate similar to double-sided sticky tape (Fig. 10.16).

Dry OCA is also available without the substrate, in which case it is just a layer of adhesive. This is also referred to as no-base OCA (Fig. 10.17).

Fig. 10.15 LCD bonded
with optically clear adhesive
(OCA)

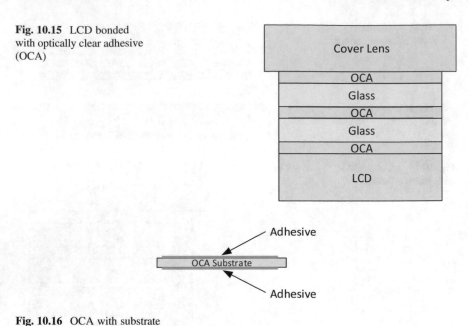

Fig. 10.16 OCA with substrate

Fig. 10.17 No-base OCA

The benefit of no-base OCA is that eliminating the plastic substrate also elimi-nates birefringence.

Applying a dry OCA requires applying pressure to remove air bubbles. It is most often used to bond touch sensor layers together and to bond the cover lens to the touch sensor. It is not typically used to bond touch sensors to displays as the LCD may be damaged in the process.

Wet OCA is a viscous material that is applied wet then cured. UV light is used to cure the material. Some liquid OCAs are also capable of self-curing which can be helpful for curing those areas where UV light cannot reach. Liquid optical bonding is most often used to bond the LCD to the touch sensor. This is a complicated process. First a dam of OCA material has to be applied and cured around the bezel of the LCD to hold the material in place. Small voids are left in the dam for air to bleed out during the next steps (Fig. 10.18).

The placement and curing of the dam is critical in order to prevent the OCA material from leaking into the backlight of the LCD and causing bright patches on the display (Fig. 10.19).

Once the dam is in place and cured, the liquid bonding agent is dispersed onto the LCD surface. Then the touch sensor and cover lens stack up is placed on top of the LCD. UV light is shined down through the viewing area of the cover lens to cure all

Fig. 10.18 Liquid OCA dams around the LCD perimeter

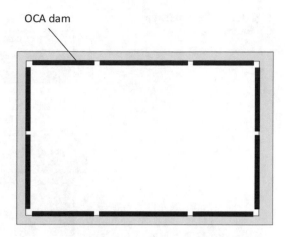

Fig. 10.19 LOB leakage

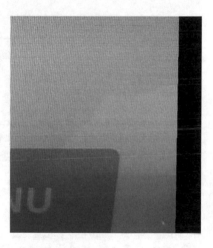

Fig. 10.20 UV curing of LOB

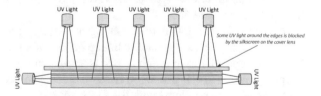

the visible material. A side firing UV light is used to cure the edges of the material (Fig. 10.20).

Properly applying and curing liquid optical bonding material is a difficult, complicated process. Each LCD is slightly different requiring a different dam

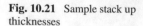

Fig. 10.21 Sample stack up thicknesses

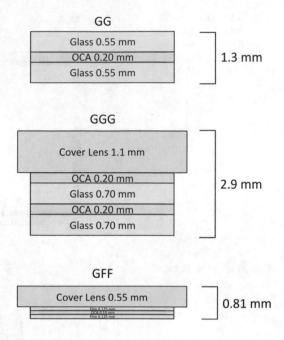

construction and curing process. And, as shown in Fig. 10.20, the silkscreen on the cover lens blocks UV light around the perimeter of the lens, preventing some areas of the LOB material from curing properly. For these reasons, it is important to find a vendor who has extensive experience with liquid optical bonding and curing. Some touch sensor manufacturers offer liquid optical bonding services, while others outsource it.

The materials used in the stack ups shown above are available in a variety of thicknesses. ITO-coated glass, for example, comes in standard thicknesses of 0.55 mm, 0.7 mm, and 1.1 mm. Dry OCAs are available in thicknesses from 10 μm to 250 μm. Some typical thicknesses of various stack up options are shown in Fig. 10.21 for reference.

We have spent a lot of time talking about the various transparent conductive materials, sensor stack ups, and optical qualities of a touch sensor. The only reason to have an optically clear touch sensor is so you can see what is behind it. The next chapter provides a basic primer on LCDs including construction, interfaces, and their potential to cause problems with the touch sensor.

Chapter 11
Display Basics

A display is referred to using the terms "LCD" and "TFT," often interchangeably. Are they the same thing? LCD stands for liquid crystal display; TFT is thin film transistor. Technically a TFT is a type of LCD. There are LCDs that do not use thin film transistors. The most common examples are segmented LCDs often seen on inexpensive home products like thermostats, microwave ovens, and washing machines (Fig. 11.1).

A full color LCD, like you see on a phone or tablet, is typically a TFT. That means it uses thin film transistors to drive the display. But drive it how? And what is a liquid crystal exactly? In this chapter we will explain what liquid crystals are, how they are used to create full color displays, what layers make up the display, and their most common physical interfaces. While there are other display types in use, TFT-based LCDs are the most common and are the only display type discussed in this chapter. For that reason, they are referred to simply as LCDs from this point forward.

The key part of an LCD is the liquid crystal. A liquid crystal is just what the name implies: a crystal that is in a semiliquid state. There are several types of liquid crystals. The type most often used in LCDs is a nematic crystal. Nematic crystals are oblong in shape and self-align so that most of the crystals are pointing in the same direction (Fig. 11.2).

Two more useful features of nematic crystals are that they can be easily forced to align to a surface material, and they will self organize into a twisted shape when placed between two layers with orthogonal orientations. If we place our liquid crystals between two polyimide films, each treated in such a way that the crystals align to those surfaces, and we rotate one of the films by 90°, we can create a twisted column of liquid crystals (Fig. 11.3).

We will show how the LCD layers are stacked as we explain the purpose of each layer. Figure 11.4 shows the first three layers.

If we shine polarized light (i.e., light whose waveforms all travel in one direction) through the stack, and the light is aligned with the bottom layer of crystals, then the

© Springer International Publishing AG, part of Springer Nature 2019
T. Gray, *Projected Capacitive Touch*, https://doi.org/10.1007/978-3-319-98392-9_11

Fig. 11.1 Segmented LCD

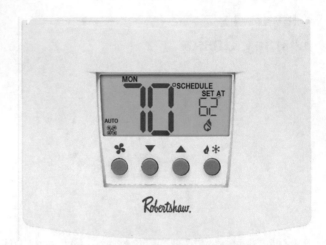

Fig. 11.2 Nematic liquid crystals

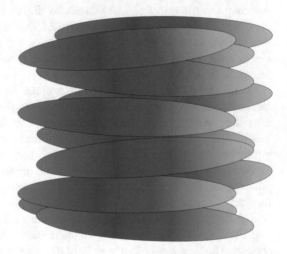

Fig. 11.3 Twisted nematic liquid crystals

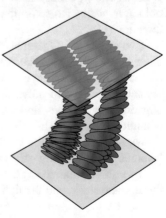

Fig. 11.4 Three display layers

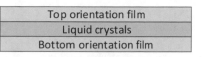

Fig. 11.5 Polarized light passing through crystal

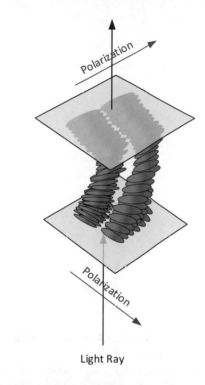

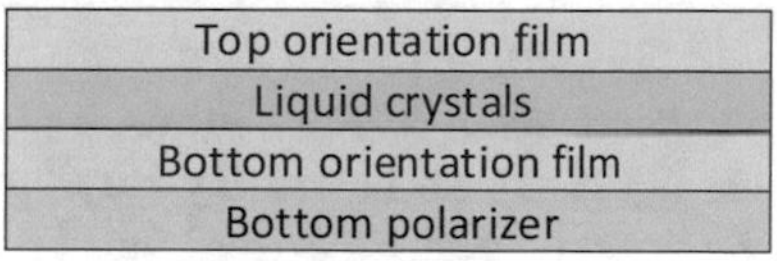

Light Ray

Fig. 11.6 Four display layers

crystals reorient the light so that it comes out at the top with cross polarization, meaning that the polarization has been rotated 90° (Fig. 11.5).

We need a polarizer underneath the bottom layer to only allow light waves with the desired orientation (Fig. 11.6).

Because of the twist in the liquid crystals, this type of display is called a twisted nematic or TN LCD. There are other types with different crystal structures, but we will not discuss them here.

Another useful feature of nematic crystals is that in an electric field the charges in the crystals organize themselves into a dipole, giving the crystal a positive end and a negative end. So in the presence of an electric field, the crystals in between the layers will try to orient themselves to the field. Because the crystals no longer maintain a smooth progression of rotations from the bottom layer to the top layer, the incident

Fig. 11.7 Untwisted
nematic crystals

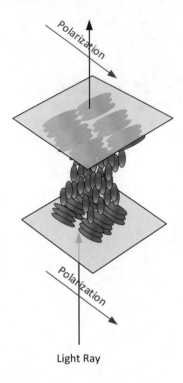

light is not reoriented and comes out the top at the same polarization direction that it
entered with (Fig. 11.7).

If we place a polarizer on top of the crystals, and that polarizer is oriented so that it
matches the orientation of the nonenergized crystals on the top layer, then the light
will pass through when there is no electric field, and it will be blocked when there is
an electric field. In other words, we can turn the light off by applying a voltage across
the liquid crystal layers. By varying the voltage between the plates, we can control
the crystals' orientation and thus how much of the light is rotated. That gives us not
just on and off but a whole range of gradations in-between. And if you know
something about electronics and you remember what TFT stands for, you are
probably already thinking that a transistor is the perfect device to control the strength
of the electric field between the plates and thus how much of the light passes through.
That is where the "thin film transistor" or TFT description comes from. The
transistors are formed in semiconductor layers sputtered onto a substrate (typically
glass) above and below the liquid crystals. The transistors themselves look a lot like
transistors formed on a standard silicon wafer (Fig. 11.8).

Note that the liquid crystals are sandwiched between the polyimide layers. The
crystals have to make contact with the polyimide film because it is the film that
defines the crystal orientation on the top and bottom layers. The semiconductor
materials that make up the transistor components are above and below the films. By

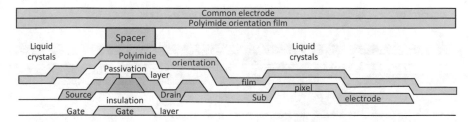

Fig. 11.8 LCD transistor

Fig. 11.9 Seven display layers

Top polarizer
Top electrodes
Top orientation film
Liquid crystals
Bottom orientation film
Bottom electrodes
Bottom polarizer

varying the amount of current going to the transistor, we can vary the amount of light that makes it through the top polarizer.

Adding the electrodes (shown as one top layer and one bottom layer for simplicity) and the top polarizer gets us to seven layers in our LCD (Fig. 11.9).

For the final step, imagine that we have three columns of crystals right next to each other. Beneath the first column is a red LED, beneath the second column is a green LED, and beneath the third column is a blue LED (LED stands for light emitting diode – it is basically a tiny colored light source). If we do not apply any current to the transistors on each column, we get the full amount of red, green, and blue light which combine to make white. If we apply the full current to the transistors, we get no red, no blue, and no green light making black. And by applying varying levels of current to each of the three transistors, we can create any combination of red, green, and blue we want, thus allowing us to create any color. The only limiting factor is how fine the current control is on our transistors. A standard TFT allows 256 levels for each transistor, giving 256 shades of red, green, and blue. That is where the acronym RGB (red/green/blue) comes from in color production. And with 256 shades for each of the three component colors, the user can make up to 16 million different color combinations. A combination of red, green, and blue transistors is called a pixel. The individual colors are referred to as sub-pixels.

Earlier we said to imagine that we have three LEDs, one for each sub-pixel. In practice, this is not economical from either a cost standpoint or a power standpoint (although this is similar to how an organic light-emitting diode or OLED display is constructed). Instead, we start with a constant white light source for all the sub-pixels. We then apply a color filter for each sub-pixel so we get red, green, and blue. The color filters are part of the electrode and film layers. There is also black mask material between each sub-pixel (Fig. 11.10).

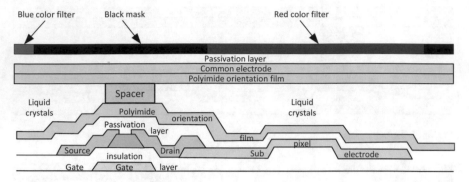

Fig. 11.10   Color filter and electrodes

Fig. 11.11   Eight display
layers

Top polarizer
Color filter
Top electrodes
Top orientation film
Liquid crystals
Bottom orientation film
Bottom electrodes
Bottom polarizer

The semiconductor traces are run underneath the black mask around and between the sub-pixels so that they do not impair the visibility of the light. The size of each pixel and the distances between them are dependent on the semiconductor process being used. The smaller the semiconductors, the higher the display resolution. Adding the color filter to our stack up gives us a total of eight layers (Fig. 11.11).

It may look odd to have the color filter above the liquid crystals, but this is where it is commonly located for ease of manufacturing. It does not matter if the color is filtered before or after it passes through the crystals. As long as we have three separately controllable sub-pixels and each of them has a color filter somewhere in the stack up, we can create whatever light combination we want.

We have left out a few layers (e.g., the electrode substrates, a light diffuser between the white light source and the bottom polarizer), and of course the layers are not to scale in the diagrams, but nonetheless the eight layers shown in Fig. 11.11 are the key layers in a TN TFT LCD. Once you understand the function of each of these eight layers, the additional layers are much easier to understand. For example, in a transflective display, some light is produced by the backlight and some is reflected light coming from outside the display. This allows the display to be seen in low light situations indoors and in high bright situations like outside on a sunny day. For this to work, several layers have to be added including wave retarders, partially reflective films, and a back reflector. But the basic operating principles of the LCD are still the same.

A typical resolution for a small LCD, say 7 inch diagonal, is 800 × 480. That is 384,000 pixels, or 1,152,000 sub-pixels. Each sub-pixel needs its own transistor with its own control lines. If we tried to drive each segment individually, we would need over 1 million sets of control lines. You may remember we had the same difficulty with trying to measure a large number of capacitors in a touch sensor. We can use the same solution here that we used for the touch sensor: drive the transistors using rows and columns. We still have a lot of control lines, but the complexity is manageable. We can also simplify things by using multiple chips in parallel to handle groups of rows and columns.

The voltages needed to drive the transistor circuits are very complicated with multiple voltage levels appearing on the same trace, forming a complex range of analog voltages. This complexity is handed by the TFT driving circuits which are part of the LCD. The key thing to understand is that there are a lot of digital and analog voltages being generated to control a large number of transistors. This results in a lot of electromagnetic interference (EMI), or radiated noise that emanates from the LCD. This noise can interfere with the operation of the touch sensor. In an earlier chapter, we mentioned that when a touch sensor element is constructed directly in the TFT cell (i.e., in-cell touch), the touch controller usually has to synchronize itself to the TFT drive signals. This is done to prevent the LCD's EMI from interfering with the touch sensor's operation. EMI can also be a problem for an out-cell touch sensor, but because of the increased distance between the transistors and the touch sensor layer, the noise does not have as much of an effect on the touch sensor measurements.

Another source of EMI noise in an LCD is the backlight. There are two major types of backlights in LCDs: cold cathode fluorescent lamp (CCFL) and light-emitting diode (LED). Like any fluorescent light source, a CCFL backlight requires high analog voltages to operate. At startup the voltage can be as high as 3000 V peak-to-peak, but it quickly drops to a few hundred volts once the lamp has warmed up. Obviously, these kinds of voltages are not provided in a normal system, so the CCFL requires an inverter. If you are not familiar with an inverter, the name is a bit misleading. An inverter is a circuit that takes a relatively low DC voltage, say 12 V, and converts it to a high AC voltage, say 120 V rms. CCFL inverters are notorious for being massive sources of EMI that can cause major problems for touch sensors. That is just one reason why CCFL backlights are becoming less and less popular on LCDs.

The more common type of backlight these days is LED. An LED can be turned on with a fairly low voltage. The intensity depends on the amount of current. In a typical LED backlight, 20 LEDs are placed in series in a string. This string is often referred to as an LED rail. Each LED has a voltage drop across it, so the more LEDs there are in the string, the higher the voltage needed to drive that string. A typical backlight voltage is 12 V. Most displays will have more than one rail in parallel, so the voltage requirement is the same but the current requirement for that voltage increases. There are also high bright backlights which require even more current.

It is possible to just turn an LED on by supplying it a constant voltage. In this case, very little EMI is generated since there are no voltage changes. If all LED

Table 11.1 LCD resolutions

Name(s)	Aspect ratio	Resolution
VGA	4:3	640 × 480
SVGA	4:3	800 × 600
XGA	4:3	1024 × 768
WXGA (Wide XGA)	16:10	1280 × 800
720p (HD)	16:9	1280 × 720
1080p (Full HD, FHD)[a]	16:9	1920 × 1080
4 k (Ultra HD, UHD)	16:9	3840 × 2160
8 K	16:9	7680 × 4320

[a]1080p is much more common now than 720p and is sometimes referred to as just "HD"

backlights operated this way, there would never be any issues with the LED rails generating enough EMI to interfere with the touch sensor. The problem is that most people want to be able to control the intensity of the backlight. The only way to do that with LEDs is to use a pulse width modulating (PWM) signal to turn them on and off very rapidly. So instead of a constant voltage the LEDs are now being driven with a pulsed waveform. That waveform can generate enough EMI to cause touch sensor problems, especially if the LCD is not well grounded. But no matter how much PWM'ing is happening, an LED backlight is still much, much quieter than a CCFL backlight. If you are selecting an LCD for a device that will also have a touch sensor, insist on an LED backlight.

There are several specifications used when comparing LCDs. The most common parameters are explained below.

Resolution is the number of vertical and horizontal pixels. There are a vast number of typical resolutions depending on the display size and aspect ratio. The most common displays today are wide aspect ratio where the ratio of width to height is 16:9 or 16:10. There are still a few sizes where the older 4:3 aspect ratio is popular (10.4 in, for example). You might expect that smaller displays have lower resolutions than larger displays, but this is not always true. Many designers are starting to require HD resolution on smaller displays. There are a variety of different standards for resolutions, and to further confuse matters, the naming is not consistent. Table 11.1 shows some typical display resolutions and the names most often used to describe them.

Brightness, as explained in an earlier chapter, refers to the amount of light coming from the display, typically measured in nits. One nit is equivalent to one candela per square meter (cd/m^2). A candela is defined in Standard Units as the amount of radiation emitted by a black body under very specific conditions. The brightness of a typical display is between 300 nits and 500 nits. High brightness displays are in the range of 800 nits to 1200 nits or more. Brightness is also called luminance in some LCD specifications.

Contrast ratio is the ratio of the brightness of the display when showing white divided by the brightness of the display when showing black. No LCD is capable of displaying true black (meaning 0 nits) when the backlight is on. There will always be some light leakage through the sub-pixels. Obviously a high contrast ratio is preferred. This can be accomplished by increasing the overall brightness (i.e., increasing the numerator of the contrast ratio), decreasing the amount of light leakage when a sub-pixel is off (i.e., decreasing the denominator of the contrast ratio), or both. The contrast ratio for a standard (meaning non-high bright) LCD is typically 400:1 to 600:1. If the brightness of a display is 500 nits when showing white, then a 400:1 contrast ratio would imply that black is about 1.25 nits. The contrast for a high bright LCD is typically 800:1 to 1100:1. That also translates to black being about 1 nit.

Color depth is the number of colors that can be displayed. This is given either as bits-per-pixel or total number of colors. The most common is 8 bits-per-pixel which yields 16 million colors. This is also referred to as true color and 24-bit color (because it is 8 bits each for red, green, and blue, and $8 \times 3 = 24$). Virtually all of the color LCD displays used in embedded devices these days are true color.

Viewing angle refers to the optimal range of angles from which to view the LCD. Within this range of angles the LCD image is clear and the colors are well defined. Outside of this range of angles the colors change or darken and the display is hard to see. There can be a bit of confusion between the viewing angle and the bias angle. The bias angle is the ideal angle from which to view a display. Typically, this bias angle is set slightly off center based on the way in which the display will be used. For example, cell phones are typically tilted away from the viewer, so their bias angle might be something like that shown in Fig. 11.12.

A wall mounted display, like an alarm panel or thermostat, is typically viewed from above, so its bias angle might be something like that shown in Fig. 11.13.

These bias angles are often referred to as 6 o'clock and 12 o'clock respectively. The viewing angle is the range around the bias angle at which the display is still acceptable ("acceptable" is often specified as "contrast ratio >10", but different manufacturers may define it however they want). For example, let's say that the LCD used in the wall-mounted panel in Fig. 11.13 has a viewing angle of 80°. The

Fig. 11.12 Cell phone bias angle

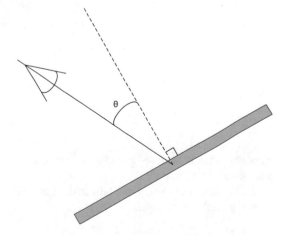

Fig. 11.13 Wall panel bias
angle

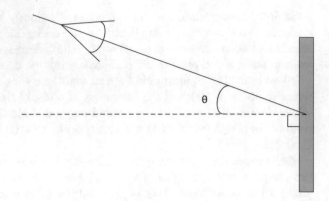

Fig. 11.14 80° viewing
angle

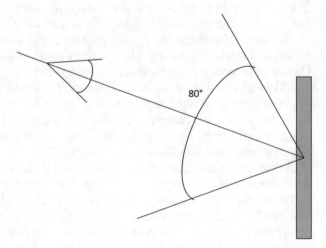

Fig. 11.15 LCD viewing
angle diagram

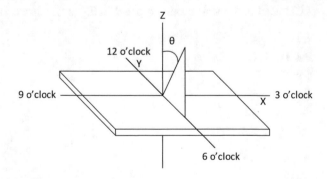

viewing angle is centered around the bias angle, so the viewing angle would be like
that shown in Fig. 11.14.

LCD datasheets often contain a diagram like Fig. 11.15 to indicate how the
viewing angle is specified.

In the case of Fig. 11.15, the viewing angle is a cone around the center of the display. In other words, the bias angle is 0°, and the viewing angle is the same from any direction. It is important to determine the bias angle as well as the viewing angle, and then consider how the user will be situated in relationship to the display. This is particularly important if the display will be rotated (i.e., used in portrait mode instead of landscape mode).

Color gamut refers to the range of colors a display can produce. It is usually specified as a percentage. The question is, a percentage of what? There are several gamut standards, but before we can explain those let's start with the basic chromaticity diagram. Human beings have three color receptors in their eyes, each tuned to a different range of wavelengths: short, medium, and long. Most humans can see wavelengths from 390 nm to 700 nm. Light can, of course, contain multiple wavelengths simultaneously, producing complex color combinations. Any color that does not contain a single wavelength is called a saturated color. Saturated colors can be broken down into their constituent primary colors: red, green, and blue. We could create a graph showing the range of visible light for all combinations of red, green, and blue, but it would be a 3D color graph which would not be much help.

In 1931, the International Commission on Illumination (CIE) created a 2D graph of the visible color spectrum. This graph has become the standard color reference when discussing the visible spectrum. The meaning of the axes and the way that specific colors are mapped into the 2D space are too complicated to explain here (and not really germane to the topic of LCDs). The graph, which has been adjusted slightly over the years, looks like that in Fig. 11.16.

The colors along the edge of outline of the shape are the single wavelength or monochromatic colors (the monochromatic wavelengths are noted around the edges

Fig. 11.16 Visible spectrum graph. (By BenRG - File:CIExy1931.svg, Public Domain, https:// commons.wikimedia.org/w/ index.php?curid=7889658)

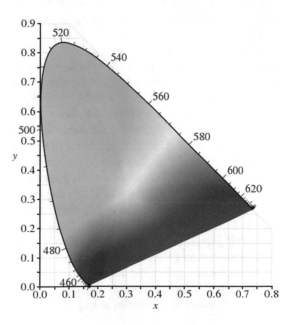

Fig. 11.17 NTSC color gamut

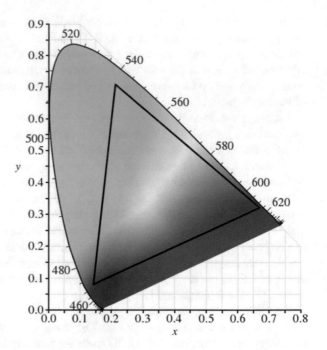

in nm). The saturated colors, or color combinations, are in the center. No LCD is capable of displaying the entire visible spectrum. Different groups and companies have defined their own color gamuts within the visible spectrum. The most popular is still NTSC, an older television standard from 1953. A device that meets the NTSC standard must be capable of displaying the range of colors within the triangle shown in Fig. 11.17.

There are other defined color gamuts including sRGB and Adobe RGB. Each gamut can be represented as a triangle within the CIE visible spectrum diagram. When an LCD datasheet states that its color gamut is 90%, it is saying that it can display 90% of the colors in a particular color gamut. The difficulty is knowing which standard is being referenced. Some LCD datasheets state that their color gamut is based on a particular standard, say NTSC. Other datasheets just give a percentage without stating a standard. Usually this means they used the NTSC standard, but not always. When in doubt, ask the LCD manufacturer for clarification.

Note that it is possible to see a color gamut specified as a percentage greater than 100%. This just means that the LCD can display more colors than are included in the color gamut's specified range. For example, if an LCD states that its color gamut is 123% of NTSC, then it can display a range of colors that lie outside of the NTSC triangle in Fig. 11.17. But if all the datasheet has is a percentage, there is no way to know what colors are included in that extra 23%.

Response time is the amount of time it takes for a pixel to turn on and off. Most display manufacturers include an explanation of how the response time is measured. A common definition is shown in Fig. 11.18:

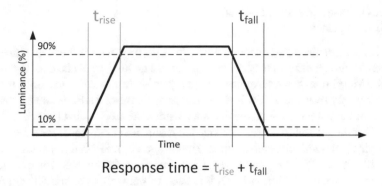

Fig. 11.18 Response time definition

A typical response time is 20–40 ms. The response time really only matters for showing video or similar applications with rapidly changing images (games for example). For most embedded devices with static screens, the response time is not an issue.

The vast majority of LCDs include some driver electronics. If you are not familiar with LCDs (as opposed to desktop monitors), then it may surprise you to learn that the interfaces to these electronics are not the consumer standards of VGA, HDMI, or display port. In fact, a desktop monitor is often just a combination of a power supply, an LCD, and a driver board that converts the external interface, VGA for example, to whatever interface the LCD requires. In embedded devices where the LCD is typically in the same enclosure as the main processor, the processor itself is often connected directly to the LCD, avoiding the need to translate from the processor's video output to HDMI, let's say, and then to the LCD's interface.

Some of the more common LCD interfaces are explained below.

Parallel or *RGB* is the simplest interface to understand. It is a parallel interface where the color data is transmitted on individual lines, one line per bit per color. For example, an LCD that supports true color (24 bit) would have 8 lines for red, 8 lines for blue, and 8 lines for green. There are also lines for horizontal synch, vertical synch, clock, VCOM (common voltage), power, ground, and a few others. Because of the large number of lines required for the color data, a parallel interface has the highest number of pins of any LCD interface. It also has the most EMI because of all the changing data on the color, clock, and synch lines. In a sense all interfaces are parallel in that the bits of the color data must be transmitted from the host to the LCD. The rest of the interfaces detailed below implement various versions of serializing the color data to reduce the number of lines, the EMI noise, and the power.

Serial Peripheral Interface (SPI) is a serial bus interface. It consists of four signals:

- SCLK: serial clock
- MOSI: master out slave in (data)

- MISO: master in slave out (data)
- SS: slave select

SPI is a master/slave architecture where the host processor is the master (meaning it generates the clock) and the LCD is the slave. The host transmits data to the slave over the MOSI line. If the slave has data to return to the master, it transmits that data simultaneously over the MISO line. A master can control multiple slaves, but each slave must have a slave select line that the master can assert so that the selected slave knows that the incoming data is intended for it and not some other slave. Depending on the physical interface (cables, wires, grounding, etc.), SPI can run at high speeds, up to 10 Mb/s. SPI is a general serial bus interface that is used for a variety of peripherals. When used for an LCD, it is used to transmit a predefined data format that is specific to graphics data.

Inter-integrated Circuit (I^2C) is another form of serial bus requiring only two signals:

- SCL: serial clock
- SDA: serial data

I^2C is also a master slave serial bus. The processor is the host and the LCD is the slave. Each I^2C packet starts with an address byte and each slave is assigned a unique slave address. All slaves monitor incoming transmission and decode the address. If the address matches the slave's address, then the slave listens to the remaining data. If the address does not match, the slave ignores the data. This allows a large number of slaves on a bus without having to add an extra control line for each device. The SDA line is duplex, meaning it is controlled by both the host and the slave, but never at the same time. If the master needs to read data from the slave, it first sends a command to the slave over the SDA line. Then the slave takes control of the SDA line and transmits the return data to the host using the host generated clock line to transmit each bit. SPI can run at speeds up to 1 Mb/s. As with SPI, I^2C is a general serial bus that is used for a variety of peripherals. When used for an LCD, it is used to transmit a predefined data format that is specific to graphics data.

Low Voltage Differential Signaling (LVDS) is a serial bus that uses differential pairs for transmitting data. Even though it is called a "voltage" interface, it is really a current interface. Data is transmitted by sending the current through the differential pair in either direction with a low voltage. When the current is running in one direction it is interpreted as a 1; when the current is running in the other direction it is interpreted as a 0. Since the pair always has current running in both directions, the electric fields (mostly) cancel each other out resulting in very low EMI noise. LVDS also requires some clocking and other control signals which are also sent in pairs. Because the current in the pairs is low (typically around 3.5 mA), LVDS is a very low power interface. LVDS can transmit up to 3 Gb/s. Technically, LVDS refers to the physical layer concept of transmitting data with differential pairs. Some LVDS interfaces can send data to multiple devices, but unlike SPI or I^2C where the host can send different data to each slave, on a multi-drop LVDS bus every device gets the same data. There is no way for the slave to send data back to the host on the same

LVDS pairs. Like SPI and I^{2}C, LVDS can be used to transmit any type of data. When used for an LCD, it is used to transmit a predefined data format that is specific to graphics data.

Mobile Industry Processor Initiative (MIPI) is class of high-speed serial bus standards designed for use in mobile systems with a variety of peripherals. There is a MIPI interface standard for cameras (MIPI-CSI), for batteries (MIP-BIF), and for several other types of peripherals. MIPI-DSI is the MIPI standard for displays. MIPI also uses twisted pairs for transmitting data (the MIPI standard refers to each twisted pair as a lane). A MIPI-DSI interface can have up to four data lanes plus one clock lane. One data lane can be reversed so that the peripheral can send data back to the host. Unlike the previous interfaces, the MIPI-DSI standard includes a specification of the graphics signals and a list of required commands that a display must support. This is a major advantage insuring compatibility between hosts and devices that support MIPI-DSI. It is also extremely high speed, capable of supporting data rates up to 6 Gb/s.

Embedded Display Port (eDP) is a variant of the display port video standard used for external monitor connections. eDP is a very full featured (and therefore complicated) specification with support for audio, backlight control, image compression, support for segmented panel architectures, and many other features. Like the other high-speed serial interfaces, eDP uses differential pairs to reduce EMI noise. eDP is capable of up to 8 Gb/s. A few years ago it seemed like eDP would supplant LVDS as the most common display interface. But to some extent the eDP standard has suffered from an excess of features making it difficult to implement and maintain backward compatibility. Several updates had to be issued to address ambiguities in the standard as more features were added. For these reasons, the MIPI-DSI interface is quickly overtaking eDP in the market.

Chapter 12
Cover Lenses, Films, and Coatings

Most products that use a capacitive touch sensor have a cover lens as the top most layer. The cover lens can be glass or plastic. If it is glass, it can be tempered or strengthened. It can have screen printed graphics. It can have holes, dead front areas, or IR lenses. It can include various enhancements like anti-glare etching or oleophobic coating. In short, there are a huge number of options when specifying the cover lens. In this chapter, we will discuss the most typical cover lens options.

We will start with the most basic option: the material. A cover lens is almost always glass or plastic. We will deal with glass first. The most common glass material is soda lime glass, also known as soda lime silica glass and float glass. This is the common, every day glass used for most applications. For cover lenses, soda lime glass is available in several standard thicknesses:

- 0.55 mm
- 0.7 mm
- 1.1 mm
- 2.0 mm
- 3.0 mm

It is possible to get soda lime glass in other thicknesses for high volume projects.

The second most common type of glass used for cover lenses is aluminosilicate glass. Aluminosilicate glass is more scratch and impact resistant than soda lime. It also has a higher resistance to thermal shock than soda lime glass.

The strength of soda lime and aluminosilicate glasses can be increased through a process known as chemical strengthening (CS). During chemical strengthening, the glass is submerged in a bath of potassium ions. As the glass soaks in the bath, sodium ions in the glass leach out through the surface and are replaced by potassium ions (Fig. 12.1).

The potassium ions are larger than the sodium ions. Inserting the larger potassium ions into the molecular structure creates internal tension (shown in Fig. 12.1 as black arrows) which increases the glass' resistance to breakage. To understand why this works, we have to understand what happens when glass breaks.

© Springer International Publishing AG, part of Springer Nature 2019   117
T. Gray, *Projected Capacitive Touch*, https://doi.org/10.1007/978-3-319-98392-9_12

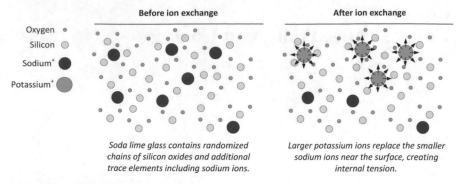

Fig. 12.1 Chemical strengthening bath

It may surprise you to learn that researchers still do not completely understand the process of how glass breaks. There are competing theories about what is actually going on at the atomic level when a crack forms and propagates in glass. Fortunately we do not have to delve to that level. At a high level the problem with glass is that it does not have an orderly crystalline structure. The molecules in glass are clumped together in chains with dangling atoms randomly distributed throughout the lattice. When the glass is scratched, i.e., when the loose lattice of molecules is damaged, that damage has a tendency to propagate through weaknesses in the lattice forming a long crack. If that crack is long enough, or if multiple cracks form from the same defect, then shards of glass will break off.

Replacing sodium ions with potassium ions forms an inner barrier to crack propagation. As long as the scratch does not go below the depth of the potassium layer, the scratch is usually contained and the glass stays together. If the scratch goes past the depth of the potassium layer, then the glass will most likely fail. So when glass is chemically strengthened, the depth of layer (DOL) is a key measurement of the process. Chemically strengthened glass with a greater DOL is generally more failure resistant than chemically strengthened glass with a smaller DOL. The DOL is typically in the range of 30–100 μm.

You may have heard of so-called super glasses like Corning's® Gorilla® Glass or Ashai's Dragontrail™ glass. These are aluminosilicate glasses that have undergone proprietary CS processes. While the details and outcomes of each process are different, they are both potassium exchange processes. Gorilla Glass and Dragtontrail have higher DOL than the standard CS process which makes them more scratch resistant. Corning and Asahi continue to improve their processes and regularly release updated versions of Gorilla and Dragontrail. You may have also heard of Sapphire cover lenses. These are cover lenses made of industrially grown sapphire. They are very scratch resistant, very expensive, only available in small sizes targeted at watches and cell phones, and are not available on the general market.

Qualifying and comparing glass strength is a difficult topic. Often people will say that a particular type of glass is stronger than another, but such statements are not really accurate since there is no single "strength" parameter for glass. The real question is how a piece of glass reacts to different types of stress. How far can you bend a piece of glass before it breaks? How much force is required to scratch it? How much force can be placed on it before it breaks? How much of an impact can it withstand before it breaks? And what if you hit it on the edge instead of on the surface? All of these questions involve different categories of breakage and are measured in different ways. The following is a brief explanation of some of the characteristics used to quantify glass.

Tensile strength measures the ability of a material to withstand a load that elongates the material. The higher the tensile strength, the harder it is to pull apart the material. It is measured as a force over a given area. The international unit of measure for tensile strength is Pascals (Pa) which is equivalent to Newtons per square meter (N/m^2). Due to the high number of Newtons required to break a material, the tensile strength is usually given in mega Pascals (MPa), which is 1,000,000 Pa. Soda lime glass has a tensile strength of 50–200 MPa. The wide range is due to the fact that scratches in the surface dramatically decrease the tensile strength. Cleaner glass with fewer scratches has a much higher tensile strength than glass with a lot of surface defects.

Compressive strength measures the ability of a material to withstand a load that compresses the material. The higher the compressive strength, the more pressing force it takes to break it. Like tensile strength, compressive strength for glass is given in MPa. Glass has a compressive strength of 600–1000 MPa.

Shear strength measures the ability of a material to withstand a shearing load. A shearing load can be thought of as two opposing forces acting along opposing planes of the material (Fig. 12.2).

Like tensile strength and compressive strength, shear strength is measured in MPa.

Modulus of rupture (MOR), also known as flexural strength, is the stress on a material just before it fails. It is typically measured using a transverse stress test where the material (in our case glass) is suspended between two points and pushed down on in the center (Fig. 12.3).

The top of the glass experiences compressive stress, while the bottom of the glass experiences tensile stress. Since glass has a much higher compressive strength than

Fig. 12.2 Shearing force

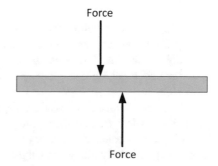

Force

Force

Fig. 12.3 Modulus of
rupture test

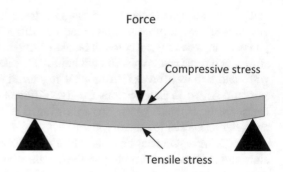

Fig. 12.4 Shear modulus

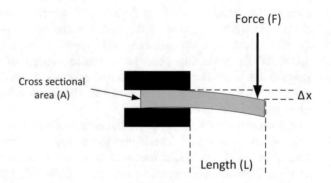

tensile strength, it fails from the bottom up. In theory, this means that the tensile
strength is also the modulus of rupture. But due to the number of surface defects on a
typical glass sample, the MOR is often less than the tensile strength. Glass typically
has an MOR of 30–60 MPa.

Shear modulus, also known as the modulus of rigidity, is a measure of the rigidity
of the material. It is measured by applying a force on the material and measuring how
far the material is deflected by that force (Fig. 12.4).

The shear modulus is calculated as follows:

$$\text{Shear modulus} = \frac{\text{Shear stress}}{\text{Shear strain}} = \frac{\frac{\text{Force}}{\text{Area}}}{\frac{\Delta x}{\text{Length}}} = \frac{FL}{A\Delta x}$$

A larger shear modulus indicates that the material is more rigid (i.e., harder to
bend). For glass, the shear modulus is typically measured in GPa (giga Pa), but note
that unlike the other forces mentioned above, the shear modulus measurement does
not involve breaking the glass, just bending it. Glass has a shear modulus of
3–5 GPa.

Young's modulus, also known as the modulus of elasticity, measures how elastic a
material is when pulled. The Young's modulus calculation is very similar to the
shear modulus calculation. The only difference is the force and deformation are
measured uniaxially along the length of the material (Fig. 12.5).

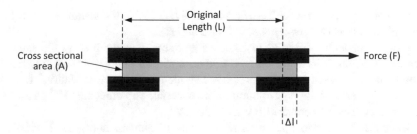

Fig. 12.5 Young's modulus

Young's modulus is calculated as follows:

$$\text{Young's modulus} = \frac{\text{Tensile stress}}{\text{Tensile strain}} = \frac{\frac{\text{Force}}{\text{Area}}}{\frac{\Delta l}{\text{Length}}} = \frac{FL}{A\Delta l}$$

A larger Young's modulus indicates that the material is not very elastic. Glass has a Young's modulus of 50–90 GPa.

Poisson's ratio is a measure of how much the thickness of a material changes as it is stretched or compressed. Think of a water balloon. If you press down on the top of the water balloon, the sides expand. If you lift a water balloon so that it stretches, the sides contract. Poisson's ratio is essentially the ratio of the how much the length of the material changes in one direction over how much it changes in the other direction. The values actually used in the ratio are strains, not lengths, but the concept is the same. The number is dimensionless. The ratio is between 0.0 and 0.5 for most materials (there are materials with negative ratios, but glass is not one of them so we will ignore this complication). The thickness of a material with a Poisson's ratio close to 0.0 barely changes at all when stretched or compressed. Cork is an example of a material with a Poisson's ratio of approximately 0.0. The thickness of a material with a Poisson's ratio of close to 0.5 thins out a lot when stretched. Rubber has a Poisson's ratio close to 0.5. Poisson's ratio for glass is 0.18–0.30.

Knoop's hardness is a measure of the hardness of brittle materials. Unlike the previous specifications, the definition of Knoop's hardness is based on a very specific test setup and process. A diamond stylus with a specific tip shape is pressed into the surface of the material at a given force and for a given time (the force and time vary for the type of material being tested). The length of the impression made in the material by the stylus is measured. Knoop's hardness is then calculated using the formula:

$$\text{Knoop's hardness} = \frac{\text{Force}}{\text{Correction Factor} * \text{Indentation Length}^2} = \frac{F}{C_P L^2}$$

The correction factor is a constant based on the geometry of the stylus. The units are typically given as HK. The test details are defined by the American Section of the

International Association for Testing Materials (ASTM) E384 standard. Glass has a Knoop's hardness of between 400 HK and 600 HK.

Vickers hardness is similar to Knoop's hardness. It is defined by the same test standard as Knoop's hardness, ASTM E384. As with Knoop's hardness, a stylus of a certain size is pressed into the material. The hardness is then calculated from the force and the area of the deformation in the material. The units are HV. Glass has a Vickers hardness between 500 HV and 600 HV.

Mohs hardness, also known as Mohs scale of mineral hardness, is similar to Knoop's hardness and Vickers hardness. The difference is the scale. The Mohs scale is based on standard minerals with talc being defined as 1 (the softest) and diamond being defined as 10 (the hardest). Being a qualitative scale, it is not as useful as Knoop's or Vickers, but it is still often listed on glass spec sheets. Untreated glass has a Mohs hardness of between 5 and 6. Chemically strengthened glass has a Mohs hardness between 7 and 8.

Dielectric constant is essentially a measure how well an electrical signal propagates through a material. The dielectric constant is dimensionless and frequency dependent. Projected capacitive touch controllers generally use signals in the hundreds of kilohertz, so we are interested in the dielectric constant of the cover lens between 100 kHz and 200 kHz. In this frequency range glass has a dielectric constant between 6.0 and 8.0.

A brief summary of each of the measurements listed above is given here along with their usefulness in qualifying cover lens materials:

- *Tensile, compressive, and shear strength*: These values can be difficult to find for some cover lens materials which makes it difficult to compare them.
- *Modulus of rupture, shear modulus, and Young's modulus*: Young's modulus is the most common of these values on cover lens material datasheets. A higher Young's modulus number is generally better.
- *Poisson's ratio:* This is a very common specification, but not very helpful when selecting cover lenses.
- *Knoop's, Vickers, and Mohs hardness:* These are one of the most important specifications for cover lens materials. But unfortunately different manufacturers use different scales which can make it difficult to compare. In any scale a higher number is better (more scratch resistant).
- *Dielectric constant:* This is also a very important specification for cover lens materials as it defines how much of the electric field makes it through the cover lens. But when it is shown on the datasheet, which is not often, it is usually given for MHz or GHz frequencies which is not helpful when evaluating PCAP performance.

In summary, you will see a lot of parameters on the datasheets for various glass and plastic cover lens materials, but comparing the values and determining which material is the best combination of price and performance for your application is difficult. Sometimes it is easier to search for an independent comparison of materials than to try to compare them yourself (i.e., soda lime vs. Gorilla glass). That being said, there are a few general rules:

1. Glass is better than plastic for scratch resistance, dielectric, and optical quality.
2. Plastic is better than glass for impact resistance and safety.
3. Chemically strengthened soda lime glass is better than standard soda lime glass for scratch resistance and impact resistance.
4. Super glasses (Dragontrail, Gorilla, etc.) are better than chemically strengthened soda lime for scratch resistance and impact resistance. They also have a higher dielectric.

So far we have only discussed glass as a cover lens material. As mentioned above, plastic can also be used. The most common plastic used for cover lenses is PMMA. Plastic has only three advantages over glass:

1. Plastic is very difficult to break.
2. When it does break, it is more safe than glass (fewer shards).
3. A clear area can be formed directly in a section of the plastic enclosure providing for a truly seamless top enclosure piece.

Plastic also has a lot of disadvantages:

1. Plastic has lower transmissivity than glass resulting in poorer optical quality.
2. Plastic scratches *much* more easily than glass (the Vickers hardness for glass is between 500 and 600; the Vickers hardness for PMMA is between 10 and 20).
3. The dielectric of plastic is about half that of glass. In order to get the same touch performance, a plastic cover lens has to be half the thickness of a glass cover lens.
4. Plastic is not very flat over large surfaces, making it difficult to optically bond to a touch sensor. The larger the screen, the more difficult bonding becomes.
5. The thermal expansion of plastic is very different from glass. When a plastic cover lens is bonded to a glass touch sensor and put through environmental stresses, failure is much more likely.

There are very few applications where plastic is the better choice, and most of those involve safety. For example, if you are developing a DVD player that is designed to hang on the headrest of a vehicle so that passengers in the backseat can view it, then the device must undergo a head impact test. During the test, a large metal hemisphere is swung down through an arc directly onto the surface of the device. In order to pass the test, there must be no loose shards of material that could cut a person's face. It is extremely difficult for any glass cover lens to survive this test. That leaves three options:

1. Use a thin protective layer over the glass to prevent the shards from escaping. This material is typically some kind of plastic film.
2. Use tempered glass.
3. Use plastic.

Tempered glass might sound like the ideal option for this application. If you are not familiar with tempered glass, this is the kind of glass used in car windshields. It undergoes a special thermal shock process which creates a large amount of inner tension. When tempered glass fails, it shatters into small, relatively harmless pieces.

Fig. 12.6 Reflected light

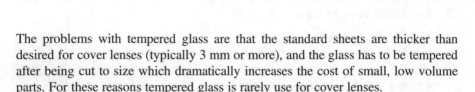

The problems with tempered glass are that the standard sheets are thicker than desired for cover lenses (typically 3 mm or more), and the glass has to be tempered after being cut to size which dramatically increases the cost of small, low volume parts. For these reasons tempered glass is rarely use for cover lenses.

Another reason to choose glass over plastic is that there are many more coatings and film options for glass. Some of the more common cover lens enhancements are discussed below.

Silkscreen printing can be applied to the back surface of the cover lens. This provides a clean look and protects the paint from wear. The silk screening can be as simple as a black border or be a complicated multicolor logo or other graphics. Special inks can be used to create dead front areas. These are areas that look solid but become semitransparent when a light is turned on behind them. One particularly useful type of dead front is an infrared (IR) window. This is a window that appears black but allows IR light to pass through. The inks are typically specified using Pantone® colors. Even though Pantone colors are very accurate, it often takes one or two rounds of prototypes to perfect the desired look.

Anti-glare (AG) reduces reflections on the surface of the cover lens. When external light hits the top surface of the lens most of the light passes through but some is reflected back (Fig. 12.6).

The reflected light causes a ghost image (i.e., reflection) to be visible on the surface of the cover lens. This ghost image can be reduced or eliminated by diffusing the reflected light. This can be done by etching the top surface to create tiny peaks and valleys; it can be done by depositing material onto the surface, also creating peaks and valleys; or it can be done by applying a film with a specialized coating that diffuses the light (Fig. 12.7).

Scattering the reflections reduces the coherence of the reflected light and mini- mizes visible reflections. AG is measured in gloss units (for more information on gloss refer to Chap. 6 "Measuring Optical Quality"). Untreated glass is typically around 140 gloss units. AG glass is between 130 and 60 gloss units. AG coatings and films can be damaged over time, so etching is the best option for AG. But etching can also be difficult or impossible if there are other required treatments. For example, etching chemically strengthened glass is a difficult, expensive process. Another issue with AG is that it not only diffuses the incoming light, it also diffuses the light

Fig. 12.7 AG etch

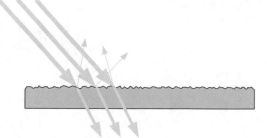

coming from the display. For these reasons AG is not often used in touch sensor applications.

Anti-reflective (AR) coatings and films are similar to AG in that they reduce reflections, but AR uses a different mechanism. The reason that light is reflected in the first place is because of a mismatch in the index of refraction. As explained in more detail in Chap. 6, every transparent material has its own index of refraction value. When light passes from one material to another, the amount of light reflected at the interface between the two materials depends on the difference between the indices of refraction. An AR coating or film is designed to minimize the reflections caused by these differences. Essentially the AR coating or film has a gradient index of refraction that slowly transitions from one value to the other. This reduces (but does not completely eliminate) the reflections. As a result, more of the incoming light is passed through the top layer of the cover lens and less is reflected back. It also means that as light from the display travels through the cover lens, less of it is reflected back toward the display as well. In other words, AR helps reduce external reflections *and* improve the quality of the displayed image. For these reasons, and because of the AG problems mentioned above, AR is much more common in touch applications than AG. AR is measured as a percentage of light reflected at the surface. Untreated glass typically has a reflection of around 4%. AR glass can have a reflection level as low as 0.5%.

Oleophobic coating, also known as anti-fingerprint and anti-smudge, is a coating on the surface of the cover lens that rejects oil molecules like those found in fingerprints. Oleophobic literally means "oil fearing." When a drop of oil is placed on an oleophobic surface, the drop will resist bonding to the molecules on the surface. In essence, the surface tension of the molecules in the drop is greater than the bonds between the oil and the coating. Instead of spreading out, the drop curls up into a nearly spherical ball (Fig. 12.8).

When a finger touches an oleophobic coating, less oils and fats are transferred from the finger to the surface of the material. And when the surface is wiped off, whatever oils are present transfer easily from the oleophobic surface to the cloth. Thus the surface tends not to collect fingerprints, and whatever fingerprints do accumulate are easily removed. The only issue with an oleophobic coating is that it is not permanent. It wears off over time through reactions with acids transferred from fingers and through erosion from cleaning.

Fig. 12.8 Oleophobic coating

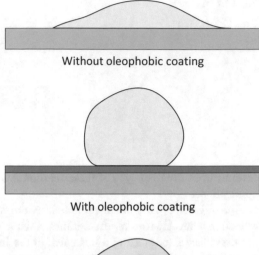

Without oleophobic coating

With oleophobic coating

Fig. 12.9 Hydrophobic coating

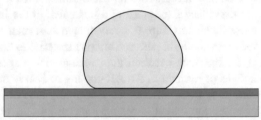

Hydrophobic coating is similar to oleophobic coating except that it acts on water (hydrophobic means "water fearing"). A surface that has been treated with a hydrophobic coating tends not to bond to water molecules. Drops of water curl up into spheres (Fig. 12.9).

Hydrophobic coatings also wear off over time. Fortunately, there is a readily available commercial product called RainX® which can be used to replenish a hydrophobic coating that has worn off. While hydrophobic and oleophobic coatings are different, they both do a good job of rejecting water and oil, they just do a little better job of rejecting one or the other. In marine and medical applications, hydrophobic coatings are very common (for more details refer to the chapter on glove and moisture performance).

There are two major parameters used when specifying an oleo- or hydrophobic coating. The first is surface energy and the second is contact angle. Surface energy is the energy associated with intermolecular bonding between materials. It is given either in Joules per square meter (J/m^2) or Newtons per meter (N/m). One Newton is one Joule per meter so J/m^2 is the same things as N/m. A high surface energy indicates that a material easily adheres to other materials. A low surface energy means that they do not. Another way to think of it is that surface energy is a measure of the number of available bonding partners. An oleo- or hydrophobic coating reduces the number of available bonds for an oil or water molecule to bind to which reduces the surface energy. Untreated glass has a surface energy between 0.03 N/m and 0.04 N/m.

Surface energy is well defined but difficult to measure. It also only takes one material into account. For example, a hydrophobic coating repels water more easily than it does oil, but the surface energy of the material does not tell you that. That is why the more common measurement for hydrophobic and oleophobic coatings is contact angle. When a drop of the liquid in question is placed on the surface, it forms one of several possible shapes. These shapes can be generalized into circular arcs (they actually tend to be a bit elliptical due to gravity, but we are ignoring that). If there is a lot of bonding between the surface and the liquid, the liquid spreads out (Fig. 12.10a). If there is a moderate amount of bonding, the liquid forms more of a hemisphere (Fig. 12.10b). And if there is very little bonding, the liquid forms a nearly complete circle (Fig. 12.10c):

These three different levels of bonding can be characterized by the angle between the surface and the contacting edge of the liquid. The surface forms one side of the angle. The tangent to the liquid at the point of contact forms the other (Fig. 12.11). That is why this measurement is called the contact angle.

The contact angle (Θ_C in Fig. 12.11) of a non-hydrophobic surface and a drop of water is typically less than 45°. The contact angle for a hydrophobic surface is typically greater than 100° (RainX on glass results in a contact angle of around 110°). Super hydrophobic surfaces can have a contact angle greater than 150°. It is easy to tell if a surface is hydrophobic. Water on a non-hydrophobic surface spreads out in connected rivulets and puddles. When you move your finger around, the water essentially smears across the surface. On a hydrophobic surface the water forms into

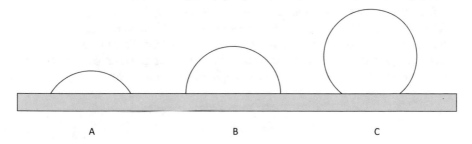

Fig. 12.10 Bonding levels

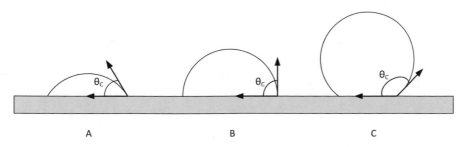

Fig. 12.11 Contact angle

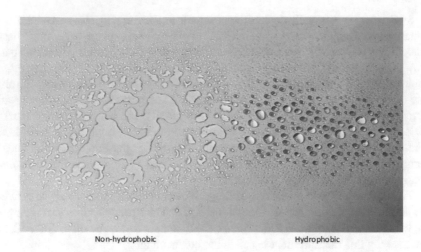

Non-hydrophobic   Hydrophobic

Fig. 12.12 Hydrophobic coating

separate balls. When you move your finger around, the water balls are pushed around the surface. Also the water tends to collect around the tip of your finger. And if the surface is tilted, the water tends to roll off. Another easy way to tell is to try to write on the surface with a Sharpie®. If the surface is hydrophobic, the ink from the Sharpie will not transfer to the glass. Figure 12.12 shows a glass cover lens which has been treated with a hydrophobic coating on the right side.

The same contact angle measurement can be performed on oleophobic surfaces with drops of oil. The contact angle ranges for oleophobic coatings are a bit smaller than hydrophobic coatings: a contact angle of less than 45° is not considered oleophobic; a standard oleophobic surface has a contact angle of around 80°; and a super oleophobic surface has a contact angle of greater than 90°.

Chapter 13
Touch Controllers

In this chapter, we will explain some of the commonalities of projected capacitive touch controllers and list some of the companies that provide them. Unfortunately a lot of the information about the actual controllers is proprietary and covered under nondisclosure agreements, so we have to limit ourselves to generalities and publicly available information. Hopefully, this chapter will give you enough information so that you can understand whatever proprietary information you receive from the touch controller vendor.

There are about two dozen companies that sell controllers targeted at projected capacitive (PCAP) sensors. Here is a list of the most well-known suppliers:

- Broadcom
- Microchip (formerly Atmel)
- EETI
- Melfas
- Synaptics
- Solomon Systec
- FocalTech
- Cypress
- Goodix
- Ilitek
- UICO
- Sitronix

Some controllers are only available for large volume applications like cell phones and tablets; some are targeted at specific devices like credit card readers; and some are only available in Asia. Many companies do not sell their touch controllers on the open market. The controllers are only available to large volume customers and design partners. The reason for this is that setting up and tuning a PCAP touch controller is a very complicated process requiring detailed knowledge of how the chip functions and what each parameter in the chip does. Companies do not generally want to provide the level of support necessary to understand their

© Springer International Publishing AG, part of Springer Nature 2019   129
T. Gray, *Projected Capacitive Touch*, https://doi.org/10.1007/978-3-319-98392-9_13

controllers to a large number of low to medium volume customers, and they do not want to make a lot of information about how their touch controller works public as this could expose some of their intellectual property. Instead, they form relationships with partners. The controller manufacturers train their partners on how to use their controllers including designing the touch sensor, laying out the PCB, and tuning the final product. These partners then provide touch sensors, controller boards, and support directly to the customers. This has been a very successful model for making touch products available to low and medium volume markets (full disclosure: my company, Dawar Technologies, is a design partner for EETI and Microchip).

Most touch controllers follow the drive/sense electrode grid concept outlined in Chap. 2. Most of them refer to the drive electrode as X and the sense electrode as Y. This is not just nomenclature. The X electrodes define the X axis coordinates, and the Y electrodes define the Y electrode coordinates. The X = dive and Y = sense architecture is so common that for the rest of this chapter we will use the terms interchangeably.

The drive signals on the X electrodes may be pulses, sine waves, or more complicated voltage patterns. The sense electrodes are almost always connected to integrators in the controllers. Some touch controllers have enough integrators to measure all the sense lines simultaneously. Other controllers, especially those designed for large sensors, have fewer integrators than sense lines. These controllers use multiplexers to measure groups of sense lines. That means they have to drive all of the X electrodes multiple times, once for each multiplexed group.

Although different controllers use different terms and different algorithms, most of them include some way of specifying the following three parameters:

1. Drive signal time
2. Sense signal gain
3. Touch threshold

The drive signal time defines how long an X electrode is energized before the sense electrodes are measured. Smaller parts with lower resistance require less time for the drive signals to propagate down the electrode; larger panels with higher resistance require more time. This is a key part of the tuning of the controller. If the drive signals are not given enough time to energize the full length of the electrode, then the edge farthest from the electrode connection point will register lower signals than the rest of the sensor. But the longer the drive electrode is energized, and therefore the longer the sense line is integrated, the more noise gets into the sense line measurement. A correctly tuned panel allows the drive signal just enough time to reach the end of the drive electrodes but no more.

The sense signal gain is the amount of internal gain on the sense measurements. If you are not familiar with gain, it is an electrical design term for multiplication. The sense signals are very low, typically less than 50 mV. To provide an adequate dynamic range for the measurement, the incoming signal is put through an amplifier with a settable gain. A higher gain gives a larger signal. It is important that the touch controller have some way of setting the gain for the sense measurements as different touch panels will have very different signal levels. And even on the same touch sensor the signal levels can be greatly affected by the thickness of the cover lens. The signal levels are also affected by environmental factors like heat and humidity, as well as by grounding. A cell phone sitting on a wooden desk generates a lower touch signal than a touch sensor in a well-grounded kiosk. Selecting the right gain has to take into consideration all possible use cases of the product. Some controllers combine the drive charge time and gain into one setting since having a longer drive and integration time means high signal levels. Other controllers provide separate parameters for the drive time, the integration time, and the gain. And still other controllers provide similar settings but with totally different definitions.

The touch threshold is a cutoff point for determining that a touch has occurred. Generally the controller provides some type of internal measurement that represents the strength of the touch signal. If the touch signal is greater than the touch threshold, then a touch is reported. If the touch signal is less than the touch threshold, then a touch is not reported. Of course, some hysteresis is also needed to ensure that a low-level signal does not constantly go in and out of touch detection.

Some controllers have very straightforward parameters that are easy to under-stand (although not always easy to balance for the best performance). Other con-troller vendors have complicated touch detection algorithms with complex parameters that are very far removed from the actual signal measurements. That is one of the reasons that touch controller vendors do not want to support smaller customers directly. Just explaining the basic tuning process can take several days of training. And some controller vendors have hundreds of other parameters to control everything from glove detection to frequency hopping to water immunity and many, many more features.

Another thing that all touch controllers have in common is a drive voltage. This is the voltage level used for the signals sent out on the drive electrodes. Some controllers use the core voltage of the controller itself (1.8 V or 3.3 V). Some controllers have built-in voltage doublers or tripplers to boost the drive voltage as high as 10 V. Some controllers use an external voltage source that can be as high as 40 V. The benefit of a high drive voltage is that it results in a higher signal level and therefore a higher signal-to-noise ratio. A higher drive voltage can also help with thicker cover lenses. This might be a little confusing. We said way back in Chap. 1 that a projected capacitive touch sensor works by measuring changes in capacitance. Capacitance is a property defined by distances and dielectrics:

$$C = \varepsilon \frac{A}{d}$$

Voltage does not appear in that equation. Increasing the voltage is not going to change the capacitance (or the difference in capacitance between no touch and touch). So why would increasing the drive voltage help us detect a touch through a thick cover lens? The reason is that we are not measuring capacitance, we are measuring voltage. The difference in the signal level between no touch and touch is relative to the change in capacitance of the sensor: the closer the finger, the greater the change in capacitance. But the signal level is also relative to the drive voltage: the greater the drive voltage, the greater the sense measurement. Increasing the drive voltage gives us a larger signal change between touch and no touch. This essentially extends the range of the electric field so that we can detect further out (or through thicker cover lenses).

When selecting a touch controller, the most important attributes are the number of X and Y electrodes. This defines the maximum size of touch sensor that the chip can support. Just take the number of X and Y electrodes, multiply them by the electrode pitch, and you can easily calculate the maximum sensor size. The trick is figuring out the right electrode pitch.

The average adult human finger is approximately 9–10 mm in diameter (we are considering the finger to be round instead of elliptical just to simplify things). A child's finger can be as small as 5 mm. It is generally a good idea for a touch to be over at least two electrodes at a time as this improves linearity and accuracy. So if we want to have excellent linearity and accuracy, and if we want our touch sensor to work for 99% of the population, then our maximum electrode pitch should be around 6 mm. There are other factors to consider that might drive us to an even lower number. For example, a fine tip stylus might require a tighter pitch (we will address this in more detail in Chap. 16). There is also the issue of finger separation. The closer two fingers are to each other, the more difficult it is to tell them apart. At some point they merge into one touch. A tighter pitch allows two fingers to be closer together and still register as two touches. For these reasons, a touch sensor that has to pass the Microsoft® Windows® Touch Certification typically has an electrode pitch between 4 mm and 5 mm.

So if 4 mm gives the best performance, why not just use that for all sensors? The reason is that a high electrode pitch requires more electrodes which means a larger touch controller which means higher cost. For comparison, let's consider a 10.4 in touch sensor. A standard 10.4 in monitor with an aspect ratio of 4:3 has an active area of approximately 228 mm by 180 mm. At an electrode pitch of 4 mm, we would need an electrode configuration of 57 by 45 or 2565 nodes (a node is an XY crossing). If the pitch is 6 mm, we need an electrode configuration of 38 by 30 or 1140 nodes. If we are willing to sacrifice some linearity and accuracy, we can go as high as an 8 mm pitch for which we only need 29 by 23 electrodes or 957 nodes. Going from a 4 mm pitch to an 8 mm pitch allowed us to reduce the number of physical connections on our controller from 102 to 52. That could reduce the controller cost by half.

Consumer products often have a very tight electrode pitch to support high accuracy drawing, dragging, pinch and zoom, etc. An industrial control that only has buttons on its user interface does not require that level of accuracy and can use a wider electrode pitch with a smaller, less expensive touch controller. This is one reason that the intended application can make a big difference in the cost of the touch sensor. This is also why it is very important to tell the touch sensor designer about the application including requirements for stylus, pinch and zoom, two finger drags (i.e., finger separation), and any required certifications.

In the discussion about the 10.4 in. you may have noticed that we did not say which numbers of electrodes were X and which were Y. Different touch controllers will have different I/O counts. Some might have more X electrodes than Y, and some might have more Y electrodes than X. Generally speaking, it is better to have more Ys than Xs. This might sound backward since it means we need more integrators. The reason has to do with the electrode resistance.

Think of a standard monitor in landscape mode. One set of electrodes runs top to bottom (vertical) and the other set runs left to right (horizontal). Because the monitor is wider than it is tall, we need more of the vertical electrodes than the horizontal electrodes. The vertical electrodes will also be shorter than the horizontal ones. Shorter electrodes have lower resistance. When we are driving an electrode with a 3.3 V signal (or 6 V or 40 V), we are pushing a lot of energy down that electrode. The resistance can be fairly high because of the comparatively large voltage. The signals on the sense electrodes, on the other hand, are in the millivolt range. We want those electrodes to have very low resistance. Therefore, we want the longer, horizontal electrodes to be X (drive) and we need the shorter, vertical electrodes to be Y (sense). This is especially true on large panels where the electrode resistance is critical. On smaller panels, the difference in resistance is not as important, which is why some controllers designed for smaller sensors will have more Xs than Ys to reduce the required number of integrators.

Other major factors to consider when selecting a touch controller are the communications interface and operating system support. We will deal with those issues in the next chapter.

Chapter 14
Communication Interfaces, Operating Systems, and Drivers

There are three primary communication interfaces used on touch controllers:

- SPI
- I^{2}C
- USB

SPI and I^{2}C, which were discussed in detail in Chap. 11 Display Basics, are master/slave serial interfaces. The touch controller is the slave and the system processor is the master. They both use an interrupt line driven by the touch controller to indicate that new data is ready. The SPI and I^{2}C standards only define the hardware interface signals and how they work. The commands used to read data and the format of the data, also called the software protocol, is different for each touch controller. Therefore, each touch controller requires a custom driver for SPI or I^{2}C. SPI and I^{2}C are generally used on embedded operating systems like VxWorks, QNX, FreeRTOS, μC/OS, and others. Touch controller vendors do not usually have drivers available for embedded operating systems. Instead the touch controller vendor will supply detailed information on the software protocol so that the end customer can write their own driver.

Universal Serial Bus or USB is also a serial bus, but unlike SPI and I^{2}C, the USB standard defines both the hardware and software protocols. USB is also a bidirectional bus so no interrupt is required. The USB software protocols define around 20 different device classes. These include audio devices, printers, USB hubs, smart card readers, and others. The most commonly used device class is human interface device or HID. This class is for keyboards, mice, joysticks, and digitizers. The HID device class is so well defined and extendable that it is often used for devices that have no human interface like uninterruptable power supplies (UPSs).

When a USB device is attached to a system, the system and the device undergo a process called enumeration. Enumeration is a complex back-and-forth stream of messages. The details are too complicated to get into here (and there are already excellent references on this process on-line and in several books). The short version is that during enumeration the device tells the host what kind of device it is, what

© Springer International Publishing AG, part of Springer Nature 2019   135
T. Gray, *Projected Capacitive Touch*, https://doi.org/10.1007/978-3-319-98392-9_14

kind of commands it supports, and what data it can provide. There is a specific digitizer device type in the USB HID specification. Within that device type are a number of what are called usages. These include pen, touch pad, white board, stereo plotter, and others. There are several different usages a touch controller could provide, which is why the USB data coming from touch controllers can differ even though they all meet the USB HID specification. The key is that the operating system includes support for USB HID. If it does, then no custom driver is needed. Just plug the touch controller into the system, wait for it to enumerate, and start using it.

One option a USB touch controller can use is to enumerate as an absolute mode mouse. This is a specific type of mouse device that provides absolute coordinates instead of relative coordinates. To understand what that means, we first have to understand how a normal or relative mode mouse works. When a mouse is moved to the right, it tells the operating system how far it moved to the right. It does not know where it is in space, just that it moved right by a certain distance. In other words, the information is relative to the original position. A touch pad is similar. If you place your finger on a touch pad, the cursor does not move until you start moving your finger around. If a touch sensor worked like this, then putting your finger on the sensor would not relocate the cursor to the finger's position. To accommodate some early digitizer pad devices, an absolute mode device type was included in the USB spec. Just like a mouse, this device type has a left button and a right button that can be up or down, but its reported coordinates are absolute. A touch is reported as a left mouse button down event, and a release is reported as a left mouse button up event. During enumeration the device tells the host what its min and max coordinates are in X and Y. It then reports touches using its native coordinate system which the USB mouse driver converts into monitor coordinates.

You might wonder why we are talking about absolute mode mice if there is an actual digitizer device type available. The reason is that some computer BIOSs include support for an absolute mode mouse but not for a digitizer. If you want to enter the BIOS and you are using a touch controller that only supports the digitizer device type, then your touch sensor may not work in the BIOS screens. That means you have to have an external mouse when accessing the BIOS. Most computers nowadays, especially those that are sold with integrated touch sensors, support the digitizer device type in their BIOS. But it is not unusual to still find older computers, particularly those that are sold as embedded platforms, that do not include this support. Some touch controllers include an absolute mouse mode option specifically for this reason.

Another communications option is something called HID-over-I^2C. This was developed for embedded systems that wanted to use the predefined digitizer device type, but did not want to dedicate a USB port to a touch sensor. One reason is that running over USB generally draws more power than running a device over SPI or I^2C. To address these concerns a special hybrid device class was developed by Microsoft. A HID-over-I^2C device uses the I^2C hardware interface (SDA, SCL, and an interrupt) and implements a specialized version of the USB HID protocol over that hardware layer. This allows touch controllers to be lower cost and lower power while still supporting a standard device driver built into the OS. HID-over-I^{2}C is a bit complicated to set up in Windows. Certain information has to be added to the Advanced Configuration and Power (ACPI) tables to define what interrupt is being used for the interface. This process is specific to a particular BIOS or, in the case of an embedded system, the board support package.

Another issue with the HID-over-I^{2}C protocol is that it supports very small data packets. This is fine for standard touch data, but when you are trying to tune the controller or debug an issue, using HID-over-I^{2}C can be painful. If HID-over-I^{2}C is required, consider implementing either a second debug bus (many touch controllers include support for a high-speed debug bus that is only used for tuning and debugging) or use a development board that can be easily switched from the standard I^{2}C protocol to HID-over-I^{2}C and back again.

Currently, HID-over-I^{2}C is used almost exclusively with Windows machines. HID-over-I^{2}C support was added in Linux kernel v3.8. But if you are using Linux, it might be easier to use standard I^{2}C. Most touch controller vendors provide an I^{2}C driver for Linux.

In the next section, we will discuss touch support in several operating systems.

Microsoft Windows XP supports single touch only. It also includes support for an absolute mode mouse. If you plug a USB touch controller into a Win XP machine, then the touch controller either needs to support an absolute mode mouse or you need a custom driver. Even though the OS itself does not support multiple touches, an individual application can support multi-touch by communicating directly with the touch controller. Of course, this means that the application has to be written for that specific touch controller.

Microsoft Windows 7 was the first version of Windows to include full multi-touch support including gesture recognition. You can plug any USB touch controller into a Win 7 machine and it will work. There is no need for a custom driver. When a USB HID digitizer is connected to a Win 7 machine, the OS provides an application called "Pen and Touch" for configuring the touch driver. To find this application just type "Touch" into the Start menu's search bar. The "Pen and Touch" app allows you to set things like the double click speed and flick settings. Another useful app is the "Tablet PC Settings" which can also be found through the search bar. This app includes a feature that allows you to assign a touch sensor to a particular monitor. Let's say, for example, that you are using two monitors, one with a touch sensor and one without. You set up the monitors next to each other left to right and tell Windows to use Extends mode. The right monitor has the touch sensor. But when you touch the sensor the cursor appears on the left monitor. How do you tell Windows that the touch sensor is actually on the right monitor? That can be done through the Setup feature of the Tablet PC app. When you start Setup, Windows places a message on the first monitor and asks you to touch it or to hit Enter to move the message to the next monitor. Using this feature you can easily define which touch sensor is associated with which monitor. The Tablet PC Settings app also includes a Calibrate feature. Generally speaking using an operating system's calibrate feature with a PCAP touch panel is a bad idea. As discussed in Chap. 1, the coordinates for a PCAP sensor are defined by the locations of the electrodes. Unlike a resistive touch sensor, no calibration is required for a PCAP sensor. There are also orientation settings in the touch controller that allow you rotate and flip the touch coordinates to any orientation. If you are having problems with accuracy or orientation, it is much better to adjust the touch controller's parameters than to use an OS calibration procedure. Oftentimes the calibration procedure can actually make things worse.

Microsoft Windows 8 added support for the HID-over-I^2C device class.

Microsoft Windows 10 supports standard HID and HID over I^2C. And of course Windows 10 is the first Microsoft operating system specifically designed for a touch interface.

Linux began supporting multi-touch in kernel v2.6.30. Support for the USB HID digitizer device class was added in kernel v2.6.38. Any touch controller that supports the USB HID digitizer interface can plug into a Linux machine running kernel v2.6.38 or later and work. Of course as is often the case with Linux, it is not always that simple. The HID drivers have to be enabled in the `kernel.config` file using the following options:

```
CONFIG_HIDRAW=y
CONFIG_USB_HIDDEV=y
CONFIG_USB_HID=y
CONFIG_HID_PID=y
CONFIG_HID_SUPPORT=y
CONFIG_HID_MULTITOUCH=y
CONFIG_HID=y
```

Alternatively, HID support can be added under the "Device Drivers, HID support" section of `make menuconfig`. If you are using some other hardware interface (SPI, I^2C, RS-232, etc.), the touch controller vendor most likely provides one or more Linux drivers to support that. The Linux driver architecture has changed several times. Make sure you get the driver that matches the kernel version you are using. While the kernel has support for touch events, it does not have native gesture recognition support. Gesture recognition is typically part of the desktop package or distribution (Ubuntu Unity, Gnome shell, etc.).

Android uses the Linux kernel at the hardware level which means that support for the USB HID digitizer is available on any Android platform based on Linux kernel v2.6.38 or later. The first Android kernel to include support for the USB HID digitizer was Ice Cream Sandwich which is based on Linux kernel v3.0.1. When using Android, it may be necessary to add a `udev` permission for your specific touch controller. Ask the touch controller manufacturer for details. If you are using a non-USB interface, the same drivers that are available for standard Linux should work in Android as well. Check with the touch controller manufacturer to make sure you get the correct driver version for the Android platform you are using and to see if any modifications are needed. One of the great advantages to using Android is that it includes a very comprehensive gesture recognition engine.

There is one last operating system feature to discuss: Wake on Touch. This is a feature that allows a touch on the sensor to wake up a computer that is in sleep or hibernate mode. Support for this feature was added to Windows 7. Enabling this feature in Windows requires some or all of the following steps:

1. The touch controller has to support this feature and it must be enabled.
2. The feature must be enabled in the device's property sheet. Open the "Device Manager," then locate the touch device under the "Human Interface Devices" section. The entry may show the name of the touch controller company, or it may just say, "HID-compliant device."

Right click on the device and select "Properties." Then click on the "Power Management" tab. Make sure the "Allow this device to wake the computer" option is selected.

On some systems this option must also be enabled in the properties for whichever USB hub the touch controller is connected to. Some trial and error may be required.

3. The USB port must remain powered during sleep modes. Enabling this can be tricky. On some systems it can be done through the Windows "Power Options." Open the "Power Options" and click on "Change plan settings."

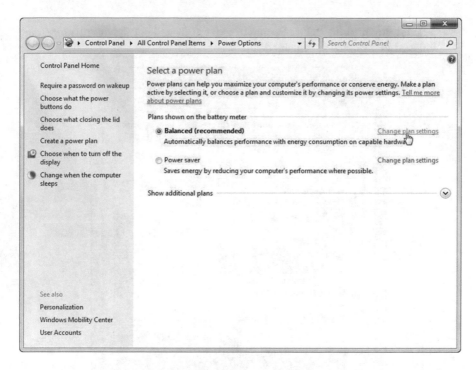

Then click on "Change advanced power settings."

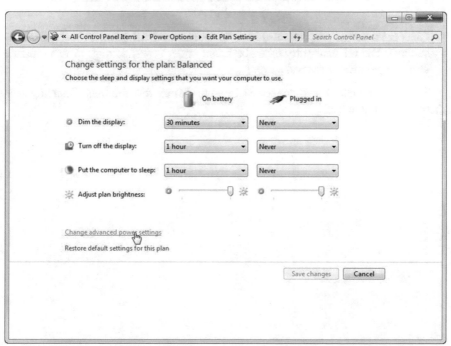

Scroll down to the "USB Settings," expand it, and set the "On battery" and "Plugged in" options to "Enabled."

4. On some systems you may also have to enable the USB ports through the BIOS settings. The relevant entry might be found under power settings or in a separate section specifically for wake events.

Wake on Touch is also supported in some Linux distributions. Enabling it is distribution dependent.

Chapter 15
Electrical Noise

Electrical noise occurs in two forms: conducted and radiated. Conducted noise is any type of electrical noise which gets coupled into a system through a physical connection. For a touch controller, the conducted noise usually appears on the power or ground connections going to the chip. Radiated noise is any type of electrical noise which gets coupled into a system through an electric field. A touch sensor is essentially an antenna that is tuned to pick up radiated signals in the 100–200 kHz range. This makes it especially vulnerable to radiated noise in the low kHz frequency range. In this chapter, we will discuss the various types of electrical noise that can affect a touch controller and sensor, the sources of noise, and ways to mitigate the impact of noise on our capacitive measurements. We will also discuss some of the standardized noise tests used for commercial and military products.

Conducted noise can sneak into a product through its incoming power or it can be created by the unit itself. A wall charger with a poorly filtered power supply is one of the most common sources of conducted noise. Figure 15.1 shows the touch response of an older model cell phone with six different chargers.

But even units with well-designed power supplies can pass noise from the external power mains into the system. In one real-world example an end user had chained together several power strips. There were multiple devices plugged into each of the power strips in the chain. A device with a touch sensor was plugged into the last power strip. The user reported numerous false touches on the touch sensor. An investigation found that one of the first power strips in the chain did not have a ground plug. Without a ground connection, all of the electrical noise that was being generated by the other devices on the power strips was distributed on the neutral and hot buses which infected the touch sensor device. As soon as the ungrounded power strip was replaced with a grounded strip, the false touches ceased.

Sometimes conducted noise appears on both the power and ground connections to the touch device. This is called common mode noise because the noise voltage level is the same on power and ground. Usually common mode noise does not generate false touches on an idle unit (meaning a unit that is not currently being touched) because the voltage difference between power and ground remains relatively

© Springer International Publishing AG, part of Springer Nature 2019   143
T. Gray, *Projected Capacitive Touch*, https://doi.org/10.1007/978-3-319-98392-9_15

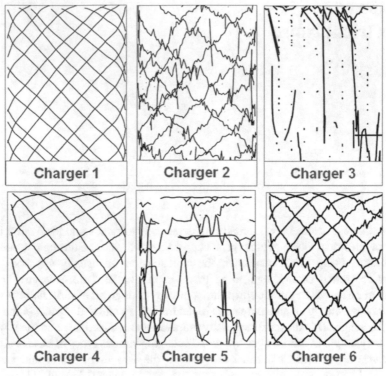

Source: Microchip

Fig. 15.1 Charger noise. (Source: Microchip)

constant. But when the user touches the sensor, they are in essence providing a steady ground to the system. The incoming noise is no longer common mode. This can cause false touches, a touch oscillating between being reported and not reported, or the location of the touch jumping around erratically. As soon as the touch is removed, the noise is common mode again and the system behaves normally. This is a common problem with devices that use Power Over Ethernet (POE). A POE power supply provides isolated power to the device without a ground connection. This is the worst case scenario: the device has a wired connection that provides a path for conducted noise but without a wired ground to shunt the noise away from the device.

The best way to deal with conducted noise is to remove it from the system. This is almost always accomplished by providing a more direct path to ground. For example, replace an ungrounded charger with a grounded charger. Make sure that the device enclosure or frame is grounded (if it is metal). If the noise is being generated within the device, separate the grounds so that the ground for the noisy components is somewhat isolated from the touch controller's ground. It might even be necessary to include some filtering between the grounds to prevent the noise from spreading through the ground planes. If the noise is on the power plane, use a linear regulator to generate the voltage used by the touch controller. In extreme cases, it may be necessary to disable the touch controller when certain system components are active.

Radiated noise is noise coupled to the touch sensor or touch controller through an electric field. The most common source of radiated noise is the LCD itself. As discussed in Chap. 11, TFTs generate a lot of radiated frequency (RF) noise. This noise passes through the touch sensor and can induce voltage changes on the sense lines. RF noise can cause problems with calibration and with normal operation. TFTs are especially prone to cause problems for touch sensors because a lot of their noise is in the low frequency range that touch sensors use. Other common sources of radiated noise are components in the system like radio frequency transmitters. These are often in a high frequency band (MHz to GHz), but they can have harmonics in the kHz range. Radiated noise can also enter the system through the touch itself. If the user is in physical contact with another device that is generating noise, that noise can actually be conducted through the body and get coupled into the sense lines of the touch sensor.

Radiated noise can be especially difficult to deal with as it cannot just be removed with better grounding. If the noise is coming from the TFT, then simply trying a different TFT might help. As mentioned in Chap. 11, switching from a TFT with a CCFL backlight to one with an LED backlight often completely eliminates false touches. Another option for limiting noise from the LCD is to add distance between the LCD and the touch sensor. This is not an ideal solution since greater distance creates more problems with parallax and generally degrades optical quality, but in some cases, this may be the only choice. Another way to deal with radiated noise is to simply over power it by using a higher drive voltage. This does not remove the noise, but it does provide a higher-signal-to-noise ratio. If the noise is coming from the LCD and the sense layer is on the bottom of the touch sensor, it might be helpful to reverse the sensor layers and put the sense layer on the top. If the noise is coming from touches and the sense layer is on top, then it might be helpful to put the sense layer on the bottom. Another option is to add filtering components to the electrode traces themselves. Resistors and ferrite beads are good options; bypass capacitors are a bad idea as they can alter the capacitance so much that the small changes in capacitance caused by a touch are no longer measureable. If the radiated noise is coming from the LCD, it can be filtered out by placing an electromagnetic (EMI) shield between the LCD and the sensor. This is usually a wire mesh layer embedded in an adhesive and with one or more external connections for grounding. An EMI shield will definitely block *any* radiated noise from the LCD, but it also adds a lot of cost and degrades the optical quality. This type of shield is more commonly used to pass a stringent radiated emissions test which is required in some applications like avionics.

If conducted or radiated noise cannot be removed from the system, then it is up to the touch controller to either filter out the noise in the measurement process or add post measurement filtering to try and remove it. One of the most common noise mitigation strategies employed by touch controllers is frequency hopping. The touch controller monitors the level of noise when no touch is present. When the noise gets beyond a certain threshold, the controller changes the frequency of the drive signal.

If the noise is still too great, the controller tries another frequency. This cycle continues until a quiet frequency is found or the controller has tried all the available frequencies. While frequency hopping is a common feature of most touch controllers, each controller implements it differently. Some controllers automatically try a variety of predefined frequencies. Other controllers allow a selection of frequencies to be specified during the tuning process. Some controllers have a highly configurable frequency hopping process; others have almost no settable parameters at all. This level of configuration flexibility is one of the things that often separates the lower cost touch controllers from the more expensive ones.

If the noise cannot be avoided by selecting a different frequency, then the touch controller might be able to filter it out of the measured signals. This can be done with a simple averaging filter, with a debounce counter (i.e., a signal must appear at the same location and with the same approximate level for a specified number of scan cycles), or with highly sophisticated algorithms. Some controllers may even use a combination of self capacitance and mutual capacitance measurements to determine if a signal is an actual touch or noise. Again, higher cost touch controllers tend to have more noise filtering algorithms with greater configuration flexibility.

As with any other system component, preventing electrical noise from interfering with the operation of the touch sensor is a matter of risk mitigation. Until a system is built and tested in an actual user environment, it is difficult to predict what the noise sources might be or what affect they will have. If you knew what kind of noise to expect and, more importantly, where it will come from, then you could address it in the design phase. This might involve redesigning the enclosure, selecting a different power supply, selecting a different TFT, placing filter components between the system and the touch sensor, etc. All of these choices are easier and less expensive to implement during design instead of after the first prototypes have been built. But at the same time you do not want to implement every possible mitigation for noise as that would overly complicate the design and drive up the cost.

The best way to mitigate the risk of noise affecting your touch sensor is to follow a few basic rules. When possible:

1. Ensure that your system uses good grounding techniques.
2. Use a TFT with an LED backlight.
3. Use a TFT with a metal back instead of a plastic one and make sure it is grounded.
4. Isolate grounds around noisy components.
5. If you are using a charger, test several models to see which is the quietest.
6. Do not place noisy components near the touch controller or the touch sensor tails.
7. Use a touch controller with a lot of configurable noise avoidance and noise cancellation features.

Perhaps the most important risk mitigation is to use an experienced touch sensor design company. A company that specializes in touch can provide guidelines for your particular application and environment. They can also review your board and enclosure design for potential issues.

There are several different standardized noise tests that are typically run on a final unit. These tests fall into three general categories:

1. Radiated emissions.
2. Radiated immunity.
3. Conducted immunity.

The first type of test, radiated emissions, is a measure of the amount of RF radiated by a device. It is intended to insure that a device will not generate enough RF to interfere with other devices in the vicinity. It is also referred to as electromagnetic compatibility testing because it tests how compatible your device is with other devices around it. The most common standard for radiated emissions is IEC 61000-6. IEC stands for International Electrotechnical Commission. The IEC standards are also known as EN standards because they are accepted by the European Committee for Electrotechnical Standardization (CENELEC). The IEC 61000-6 specification has several subparts which list emission standards for a variety of environments including residential, industrial, power stations, etc.

The only issue with a radiated emissions test and a touch sensor is that the sensor itself is an RF radiator. Depending on the type of emissions test required, a touch sensor might put out enough RF to fail the test. There are a couple of ways to reduce the emitted RF. One is to lower the drive voltage and another is to add some slewing to the drive signal (slew is an electrical term for slowing down the rise and fall times of a voltage). Slewing can be done by adding resistors to the drive lines. Some touch controllers have a slew setting that can be used to slow down the voltage transitions.

The most common standard for radiated immunity tests is IEC 61000-4-3. IEC 61000-4-3 verifies that a device works correctly when bombarded with RF energy. The test is conducted by setting up the device under test (DUT) in an anechoic chamber. There are two types of anechoic chambers: one designed to completely absorb sounds waves and one designed to completely absorb electromagnetic waves. Here we are talking about the latter. A large antenna is also set up in the room and is pointed at the DUT. The antenna is connected to a signal generator and a power amplifier in a separate room (Fig. 15.2).

The frequency generator is set up to sweep a range of frequencies from 80 MHz to 1 GHz. There are four signal strengths given in the standard which are described as test levels 1, 2, 3, and 4. The test level selected depends on the environment in which the DUT will be used. Touch devices are typically used in a class 2 or class 3 environment. For class 2, the signal strength is 3 V/m; for class 3, the signal strength is 10 V/m. A touch sensor can usually pass a 3 V/m (class 2) radiated immunity test without difficulty. Passing at 10 V/m (class 3) can be difficult depending on the controller. One of the best ways to pass a class 3 test is to increase the drive voltage.

One interesting thing about the IEC 61000-4-3 (and other IEC tests) is that it does not define the passing criteria for a device. That is up to the designer. For touch-enabled products, there are several ways that a pass might be defined in increasing order of difficulty:

1. No false touches
2. Responds to touches but performance may be degraded (does not respond to 100% of touches, see occasional false touches, etc.)

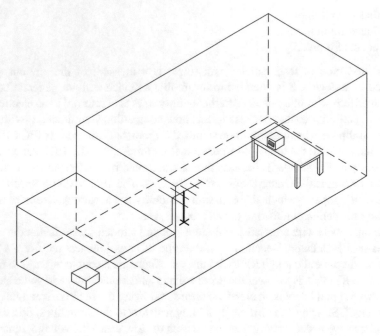

Fig. 15.2   IEC 61000-4-3 test chamber

3. Responds to touches in a normal fashion.

Of course in order to pass 2 or 3 a touch must occur during the test. A human cannot be in the chamber during the test, so this is often implemented with some kind of mechanically actuated stylus.

As mentioned earlier, one of the most common noise avoidance mechanisms in touch controllers is frequency hopping. In a normal environment, frequency hopping can be very useful since the frequency of the noise seldom changes. But during an IEC 61000-4-3 test frequency hopping can actually create a problem. The issue is with the speed of the frequency sweep. Let's say that the sweep hits a frequency that begins generating enough noise to trigger a frequency hop. The touch controller changes frequencies and conducts a noise scan to ensure that the new frequency is OK. By the time that happens the sweep has moved on to a new frequency. It is possible that the new frequency is generating enough noise that the controller tries to hop again. In theory, this loop could continue until the sweep is completed. The good news is that, if the frequency hopping is set up correctly, there should not be any false touches generated during the test. The bad news is that the controller is so busy searching for a clean frequency that it stops responding to touches. If the DUT is set up to test for touch actuation, it will most likely fail. This can be addressed somewhat by slowing down the sweep so that the controller has time to hop to a new frequency, but of course this prolongs the test. In some cases, it might actually be easier to pass the test with frequency hopping disabled.

The most common standard for conducted immunity tests is IEC 61000-4-6. This test verifies that a device works correctly when noise is present on the power going to the device. The noise is injected into the DUT's power lines using an injection clamp. The clamp is a conductive cylinder that literally clamps over the power lines going to the DUT. The clamp is connected to a signal generator. A sine wave is injected into the clamp which is then inductively coupled onto the power lines. The signal generator sweeps through a range of frequencies, typically 150 kHz to 80 Mhz. As with IEC 61000-4-3, there are four test levels defined. The test level for class 2 is an injected sine wave at a level of 3 Vrms. Class 3 requires a sine wave at 10 Vrms.

Unlike the radiated immunity test which injects noise directly into the touch sensor, a conducted immunity test injects noise into the device. If the device has good filtering and is well grounded, that noise may never even reach the touch controller. Obviously this is the best scenario since it removes any need for risk mitigation by the touch controller. If some of that conducted noise does make it to the touch controller, then the controller can attempt to avoid the noise using frequency hopping or by removing the noise with filtering or other techniques. The key point is that, once the noise makes it into the touch controller, it mostly does not matter whether it made through the sense lines or through the power and/or ground lines. Either way the noise infects the sense measurements and has to be either avoided or removed. At this point, most of the same techniques and algorithms used by touch controllers to deal with noise can be used for both radiated as well as conducted noise.

There are several other electrical tests that a touch device might be subjected to. Most of them have more to do with testing the DUT than the touch sensor. One exception to that is an electrostatic discharge (ESD) test. The standard ESD test is IEC 61000-4-2. An ESD test verifies that a device is not affected by the kind of static shock that can occur when a human touches a conductive surface. ESD testing is performed using a device called an ESD gun (Fig. 15.3).

Fig. 15.3 ESD gun

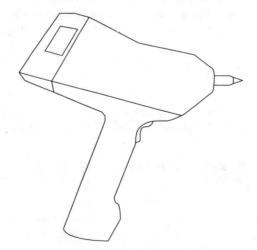

An ESD gun discharges a defined amount of static shock through a metal tip. The device can be set up to discharge when brought into contact with a conductive surface (this is called a contact discharge) or to discharge whenever the trigger is pulled (this is called an air discharge). As with IEC 61000-4-3 and IEC61000-4-6, there are four test levels defined in IEC61000-4-2. Each level has its own contact and air discharge voltages. For class 2, the contact and air discharge are both ±4 kV. For class 3, the contact discharge is ±6 kV and the air discharge is ±8 kV.

It can be surprisingly difficult to perform a full ESD test on a touch-enabled device. If the top surface is a glass cover lens and the surrounding bezel is plastic, then just getting the ESD gun to execute a contact discharge can be difficult if not impossible. Often only the air discharge test can be performed.

ESD does not generally cause an issue for touch sensor performance. First, there should be a ground ring around the touch sensor to shunt any ESD discharge straight to ground. Second, even if some of the ESD energy does get into the touch controller, it is such a brief pulse of noise that even the mildest software filtering will remove it.

So far we have only discussed commercial standards. There are also military equivalent test standards for radiated emissions, conducted immunity, and radiated immunity. We are only going to discuss the US military standards just to keep things simple. Interestingly, there is no US military equivalent of the IEC 61000-4-2 ESD test.

All of these standards fall under MIL-STD-461E. The equivalent test standards are shown in Table 15.1.

The 461E standards are quite a bit more complicated than the commercial IEC standards. The biggest complication is that 461E calls out different standards for different platforms including surface ships, submarines, Navy aircraft, Air Force aircraft, etc. Within each of those standards, there are often different test levels called out for specific areas on those platforms such as aircraft external, aircraft internal, ship above decks, ships below decks, etc. The 461E standards also tend to be more stringent and thus more difficult to pass than the commercial standards. For example, the 461E radiated frequency test range is 2 MHz to 40 GHz compared to the commercial range of 80 MHz to 1 GHz (IEC 61000-4-3).

Passing the military standards with a touch-enabled product is not inherently different from passing the commercial standards. The same mitigations come into play. You just need to be more careful about selecting the right design options in the beginning of the project. You might get away with a 10 V drive signal and pass IEC 61000-4-3 class 3, but you may have to use a 40 V drive signal to get the same product through MIL-STD-461E RS103. Identifying the required standards up front

Table 15.1 Military test standards

Test description	IEC	461E
ESD	61000-4-2	None
Radiated emissions	61000-6	RE102
Radiated immunity	61000-4-3	RS103
Conducted immunity	61000-4-6	CS114

can often drive the touch controller selection process. You do not want to get your first prototypes for a product that has to pass MIL-STD-461E RS103 and then find out that the maximum drive voltage your touch controller supports is 3.3 V.

Chapter 16
Stylus and Glove

The users of touch-enabled devices often wear gloves. These could be very thin latex gloves in an operating room or very thick parka-style gloves at an ATM in Alaska. Wearing a glove is not a problem with several touch technologies like resistive and infra-red, but it can create difficulties for PCAP. A PCAP sensor detects when a grounded object (like a finger) enters an electric field. The field typically extends only a few millimeters above the top surface. When a user is wearing a glove, their finger is prevented from reaching the surface. Essentially the glove creates a gap between the finger and the sensor. The resulting performance is basically the same as if the finger was hovering several millimeters off of the surface.

It might be reasonable to think that the glove material matters. Maybe an electric field propagates better through leather than cotton. In fact the dielectric of the material, which in essence measures how well an electromagnetic field propagates through it, makes little difference. The key parameters are the thickness of the material and how well it fits. The thinner the material, the closer the finger is to the surface. If the hand in the glove is too small and the finger does not reach the end of the finger tip, it does not really matter how thick the material is. You can always push down harder and crumple the finger tip, but most of the time the glove material just folds under itself creating an even thicker barrier.

There are gloves designed specifically for PCAP touch sensors. These gloves have some kind of electrically conductive material that connects the user's finger (or the entire hand) to the outside of the finger tip (or the entire glove). These types of gloves work fairly well and should definitely be recommended to users. But unfortunately product designers cannot always dictate the types of gloves a user is allowed to have.

Figure 16.1 shows the relative mutual capacitance signal levels of several types of gloves. The signal levels are normalized to a bare finger which is shown as 100%.

As you can see from the graph, latex gloves have little impact on the signal level. Even a double layer of latex gloves will generally work just fine without any customized tuning. The reasons are that latex is very thin and it is form fitting,

© Springer International Publishing AG, part of Springer Nature 2019
T. Gray, *Projected Capacitive Touch*, https://doi.org/10.1007/978-3-319-98392-9_16

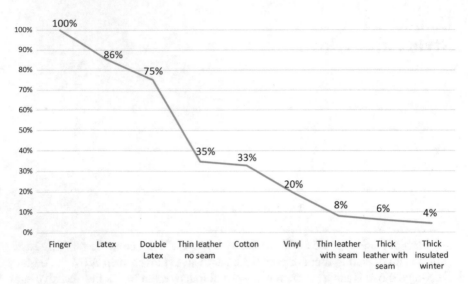

Fig. 16.1 Glove touch mutual signal levels

allowing the finger to be as close to the surface as possible, separated only by the thickness of the latex. The rest of the gloves are roughly in the order of their material thickness. There are a few exceptions for gloves that have a seam along the fingertip. These types of gloves, leather construction gloves for example, are particularly difficult because the seam adds a hard barrier between the finger and the touch surface.

The measurements shown in Fig. 16.1 are based on a mutual capacitance scan. Back in Chap. 3 we explained that self capacitance produces a bigger electric field. So how do the same gloves perform in a self capacitance scan? That data is shown in Fig. 16.2.

The mutual and self capacitance levels are shown side-by-side in Table 16.1 and Fig. 16.3.

The self capacitance signals (as compared to a bare finger) are larger for each glove type than the mutual capacitance signals. That is one reason why self capacitance is better than mutual at detecting glove touches. Another reason, not apparent when looking at relative data, is that the self signals are larger than the mutual signals giving more of a dynamic range. For the controller used to capture the above data, the self signal measurement for a bare finger was almost twice the mutual measurement. It is for these reasons that some controllers use a combination of mutual and self capacitance. Mutual is used to detect bare fingers and self is used to detect glove touches. Also note the large difference between double latex and the next glove. In the mutual data, there is a 40% difference between the double latex glove and the thin leather glove with no seam; in the self data the difference is 48%. Because the latex and double latex signal levels are much closer to a bare finger signal level than to a gloved touch, detecting a touch through a latex glove is not difficult and can often be done with just mutual capacitance data.

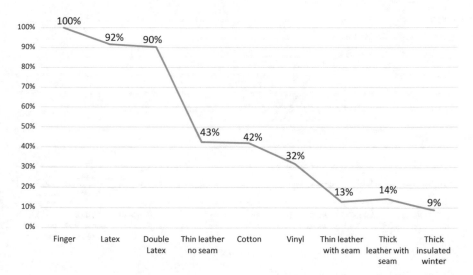

Fig. 16.2 Glove touch self signal levels

Table 16.1 Mutual and self for gloves

Glove type	Mutual Signal %	Self signal %
Finger	100	100
Latex	86	92
Double latex	75	90
Thin leather no seam	35	43
Cotton	33	42
Vinyl	20	32
Thin leather with seam	8	13
Thick leather with seam	6	14
Thick insulated winter	4	9

Some users like to use a stylus with their touch-enabled products. A stylus enables finer control and smaller touch areas. There are three types of stylus:

1. Passive untethered
2. Passive tethered
3. Active

A passive untethered stylus is the most common. The stylus is metal or metalized plastic and often has a foam tip. It is not connected to the device (i.e., untethered). If you connect the passive stylus to the ground plane of the device with a conductive wire, it becomes a tethered stylus. This type of stylus is most commonly used with credit card terminals. Grounding the stylus gives a much stronger signal. An active stylus has active, powered electronics inside of it. There are several types of active stylus. The most common emits some kind of signal which is picked up by the touch sensor or by a separate layer underneath the sensor. This signal can be used just to

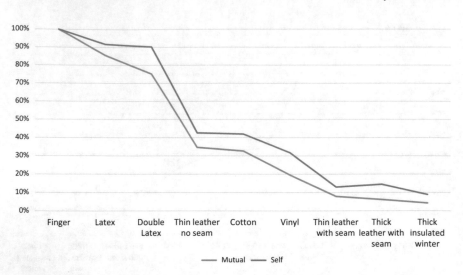

Fig. 16.3   Mutual and self for gloves

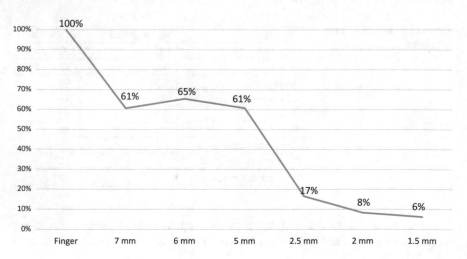

Fig. 16.4   Stylus mutual signal levels

detect location, or it can include an actual communications channel. The active stylus can use the communications bus to indicate certain events like a button press. An active stylus can have multiple buttons to execute functions like select, erase, change color, change width, etc. Active styli are available on some tablets and notebooks, but they are not very common due to the expense of the pen itself which often gets lost and has to be replaced.

Figure 16.4 shows the mutual signal levels for several untethered passive styli compared to a bare finger.

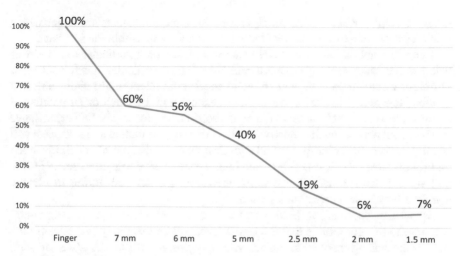

Fig. 16.5 Stylus self signal levels

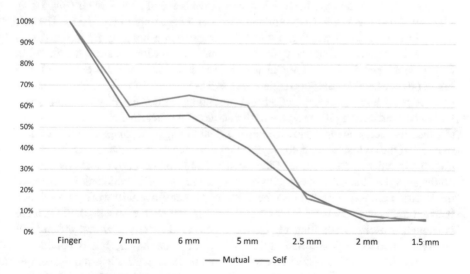

Fig. 16.6 Mutual and self for styli

The signal levels roughly correlate to the size of the stylus tips, but not entirely. This is due to differences in the materials and coatings on the stylus. Figure 16.5 shows the self capacitance signals for the styli.

Both mutual and self are shown in Fig. 16.6.

Interestingly, the relative signal levels of the styli do not follow the same pattern as the glove signal levels. Sometimes the mutual signal levels are actually higher than the self signal levels. There are several potential reasons for this including the angle of the hand that was holding the stylus and how far down the hand was on the stylus. But the key point is that, while self can also be useful for detecting a stylus, it is not as helpful as it is for detecting gloves. Still, the self signals are larger which provides more of a dynamic range for identifying a stylus within the self data.

Keep in mind that these measurements were taken on a particular touch sensor using a particular electrode pattern and mated to a particular touch controller. Different electrode patterns and different touch controllers might show different results. Regardless, the key point is that detecting a glove touch or an untethered stylus touch is difficult because of the rapid drop in the signal levels. Self capacitance can help, but only so much. Touch controllers that support mutual and self scans will usually perform better for gloves and styli than mutual only controllers. Some touch controllers have specific algorithms and tuning parameters designed to detect gloves and styli. These specialized algorithms work much better than trying to use the standard finger touch detection algorithms.

So what happens if we hold a stylus in a gloved hand? The most common response to this question is that the signal level will be extremely low because the hand is not making conductive contact with the stylus. So let's test that (we will stick to self measurements since that is where we will see the biggest signal levels). The self signal level for a finger in a thick winter glove is 9% of a bare finger (Fig. 16.2). The self signal level of a 7 mm stylus is 55% of a bare finger (Fig. 16.5). The self signal level for a hand wearing a thick winter glove and holding a 7 mm stylus is 40%. So the signal level of the stylus only dropped 15% when wearing the glove! In fact we got a much stronger signal holding the stylus than we did for the gloved finger (40% vs. 9%). This surprising result is due to several factors. First, the stylus itself is providing some level of virtual grounding within the electric field. Even without a hand holding it, the stylus pulls enough of the electric field into it to change the measured capacitance. How much the field changes depends on the size and material of the stylus. Second, the stylus does not have to be conductively coupled to the hand in order to divert some of the electric field into the body. If a gloved hand is holding a stylus, the hand is capacitively coupled to the stylus which allows some of the electric field to get drawn into the hand. A thinner glove will work better than a thicker one because the hand is closer to the stylus, but even with a thick glove there is enough capacitive coupling to provide a path to the body's virtual ground. Of course a conductively coupled stylus will work even better, but a conductive connection between the hand and the stylus is not necessary. This is just one example of how the behavior of a projected capacitive touch sensor is not always obvious.

Chapter 17
Calibration, Retransmission, and Suppression

Many people are familiar with the process of calibrating a touch sensor. What they do not often realize is that only resistive sensors need to be calibrated by the user. Calibration is required on a resistive sensor because the system needs to determine how much the resistance varies across the sensor. And since the resistance can change over time, calibration often has to be repeated.

A projected capacitive sensor does not need the kind of calibration that a resistive sensor needs, but it does still need calibration. The key difference is that a PCAP sensor self calibrates.

When a PCAP touch controller powers up, one of the first things it does is execute a calibration scan. During calibration the controller performs the same scan that it does later on when looking for a finger. The difference is that during calibration the controller stores the measured signal level for each node (a node is the crossing of an X and Y electrode). During a touch detection scan it creates a new table of signal measurements for each node, then subtracts that table from the calibration table. The difference at each node is then used to determine if a touch is occurring. Note that this is how a majority of PCAP touch controllers function, but not all. Some controllers have specialized measurement algorithms that behave very differently. Since those are proprietary, we will stick to the more common and more easily explained measurement processes.

Once the calibration data is saved, the touch controller will often adjust it slowly over time since the measurements can change slightly due to environmental changes. Of course these adjustments have to be made when no touch is present. Otherwise the controller would include a touch signal in the calibration data.

But what happens if a touch is present during calibration? If a touch gets calibrated in, then the affected nodes would have a very different signal level from the rest of the sensor. Once the finger is removed, the signals move in the opposite direction of a touch which would not generally cause any problems. But if the user touches that same area again, the signals will return to the calibrated levels. The difference between the real-time signals and the calibrated signals would be near 0 so

© Springer International Publishing AG, part of Springer Nature 2019

T. Gray, *Projected Capacitive Touch*, https://doi.org/10.1007/978-3-319-98392-9_17

no touch would be reported. In other words, touching the sensor during calibration can cause major problems.

Fortunately, it is possible to detect a touch during calibration. The measured signal levels of each node during calibration will be relatively close to each other. There will be some variance, particularly at the ends of the electrodes furthest from the sensor tails, but the differences from one node to its adjoining neighbors will be pretty small. But if a touch is present, then the signal levels of the affected nodes will be *very* different from the rest of the sensor. It is possible for a touch controller to analyze the calibration data, detect nodes with high signal levels, and quarantine those nodes. The controller can then enter a state where it constantly scans the sensor and watches those quarantined nodes. Once the signals on those nodes is more consistent with the rest of the sensor, the touch controller can execute a new calibration cycle and store the corrected signal levels. And while it is waiting for the touch to be removed, it can still scan and report touches on those nodes that were not quarantined. The detection and proper handling of a touch during calibration is another feature that often separates the more expensive controllers from the less expensive options.

What if an entire palm is on the sensor during calibration? Placing an entire hand on the sensor creates a much more complicated pattern of signals. There will of course be large signals that look like touches, but there will also be large anti-touch signals. These are signal peaks that occur in the opposite direction from the signal of a touch. If a touch signal is a positive peak, then an anti-touch signal is a negative peak. But how can a touch result in a negative peak? This occurs because of an interesting phenomena known as retransmission.

Imagine that there are two fingers on the sensor and that they are the thumb and forefinger of the same hand. Let's also imagine that the fingers are diagonal to each other relative to the sensor's orientation. The thumb is at node X2:Y4 and the forefinger is at node X8:Y13 (Fig. 17.1).

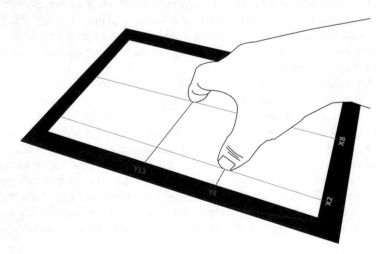

Fig. 17.1 Two simultaneous touches

Fig. 17.2 Signals for touch
at X8:Y13

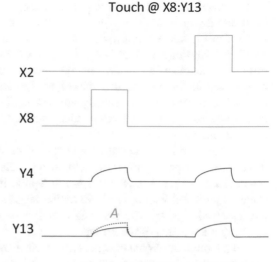

Touch @ X8:Y13

Fig. 17.3 Signals for two
touches

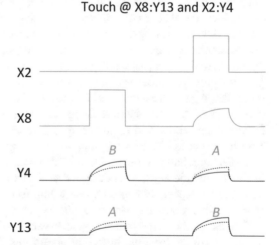

Touch @ X8:Y13 and X2:Y4

If only the forefinger were touching, then the signals on the four electrodes would look something like Fig. 17.2.

The signal level at A is slightly lower than the original signal (represented by the dashed line). The lower signal indicates that a touch is present. And since that lower signal occurs on Y13 at the same time that X8 is being pulsed, then the touch is at X8:Y13. All the other Y pulses are the same size as their original, calibrated values so no other touches are present.

Now let's look at the signals for the two touch case shown in Fig. 17.1 where one touch is at X8:Y13 and the other touch is at X2:Y4 (Fig. 17.3).

The signals at the two points marked A show a slight decrease indicating touches. These two decreases occur on Y13 when X8 is being pulsed and on Y4 when X2 is being pulsed, so the touches are at X8:Y13 and X2:Y4.

The problem is that the drive pulses of each X electrode are passed through the hand and injected into the touch panel at the other finger's location. When X2 is being driven, the X2 pulse travels into the thumb, through the hand, out the forefinger, and into Y13. When X8 is being driven, the X8 pulse travels into the forefinger, through the hand, out the thumb, and into Y4. This additional and unexpected energy on the two sense lines results in higher signal levels than the original, calibrated values. These points are marked B in Fig. 17.3. A lower signal level (compared to the calibrated value) indicates a touch, so a higher signal level indicates an anti-touch.

When a palm is on the sensor, a chaotic system of retransmission occurs. Every drive pulse that passes under the palm also travels through the palm and into every sense electrode being touched. It is impossible to predict which nodes will have lower signals and which will have higher signals due to the chaotic nature of the retransmitted signals. The result is a confusion of touch and anti-touch signals covering a large area of the sensor.

Even though the signal pattern is not as clean as a single finger, it is still possible to analyze the calibration data and determine, based on the complex pattern of touch and anti-touch, that a palm is present during calibration. If the touch controller realizes this, it can refuse to qualify any touches until the palm is removed and a normal calibration can occur. This same analysis can also be used during normal operation to detect a palm and prevent touches from being reported. This is called palm suppression since the touches from the palm are being suppressed by the touch controller. Again, these features are often available only on higher cost controllers. If a controller does not support finger or palm detection during calibration, then touching the sensor during power up can result in highly unpredictable behavior. And if the controller does not support palm suppression during normal operation, then placing a hand on the sensor will also result in erratic behavior. Some controllers even include algorithms designed to detect and remove retransmission effects. This can be especially helpful when trying to detect water. Refer to Chap. 18 for more details on how a sensor responds to water.

There are other types of touch suppression in addition to palm suppression. Some touch controllers have ear detection algorithms. These algorithms are specifically designed to detect the oblong elliptical shape of an ear so that, when a cell phone is placed against the user's head, all touches can be suppressed. Another type of suppression useful on cell phones is grip suppression. Grip suppression activates when multiple touches occur nearly simultaneously along opposite edges of the sensor. Another common suppression mode is to suppress all touches when a large number of touches are detected. This can help prevent false touches caused by electrical noise. There are also palm suppression modes specific to stylus detection so that if a user is resting their curled up hand on one part of the sensor while holding a stylus that is touching another part of the sensor, the controller can correctly suppress the touches caused by the heel of the hand while allowing stylus touches in another area. As with palm suppression, these types of suppression features are typically available on more expensive controllers.

Chapter 18
Water Immunity

Water immunity is needed in a variety of touch applications including marine devices (lake water and salt water), medical terminals (blood and saline products), exercise (drinking water and sweat), and outdoors (rain water). The most important performance criterion for water immunity is no false touches. Most applications also require some level of touch response with water. In the simplest case, a single touch response is required, for example, tapping a button. The next level of difficulty is tracking a single finger, for example, dragging a slider. Next are two-finger gestures like pinch, zoom, and rotate. The most difficult would be tracking three or more fingers. Fortunately, tracking more than two fingers is rarely a requirement on these types of devices.

Water immunity can be broadly categorized into two types: plain water (or tap water) and salt water. Plain water refers to any water that has low salinity compared to salt water. Salt water is worse because it is much more conductive. In this context "salt" refers to any compound ions that can dissolve in water and conduct a current including sodium chloride, magnesium sulfate, and others.

Water conductivity is measured in siemens per meter (s/m). A siemen is the reciprocal of Ohms ($s = 1/\Omega$). It is also sometimes called a Mho. Table 18.1 gives the conductivity for several types of water at room temperature.

As mentioned above, the concentration of salt in water is one of the best indicators of conductivity. As the salinity increases, the conductivity increases. Tap water has a salinity of about 0.15%. Sea water has a salinity of about 3.5%. The conductivity of the water (or any other liquid) is the key parameter when studying the effects of water on a projected capacitive touch sensor.

How does water cause false touches? Through the same retransmission process mentioned in the previous chapter. Imagine that we have a touch sensor lying flat on a surface. There is a puddle of water on the sensor large enough to cover four nodes (Fig. 18.1).

The water will conduct the drive signals from the X lines into the Y lines, causing the signal levels on the Y lines to increase. This is the same retransmission effect that occurs with two diagonally opposed fingers. The increased signals on the Y lines

T. Gray, *Projected Capacitive Touch*, https://doi.org/10.1007/978-3-319-98392-9_18

Table 18.1 Water conductivity

Type of water	Conductivity (s/m)
Distilled water	0.00005–0.000530
Tap water	0.005–0.080
Fresh water stream	0.010–0.200
Sea water	5.0–5.5
Sweat	5.5–16.0

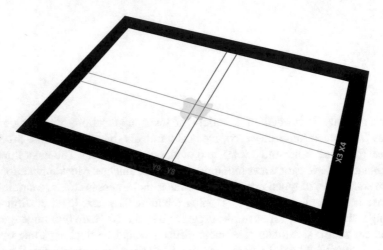

Fig. 18.1 Water covering four nodes

look like anti-touches (for an explanation of anti-touch refer to the previous chapter). The greater the conductivity of the water, the more energy gets transferred and the larger the anti-touches.

But that is not all that happens. When X3 is being driven, X4 is grounded. So the water also creates a conductive path to ground for the drive signal on X3. That lowers the signal levels on the two Y lines which looks like a touch. It would be great if the anti-touch and the touch signals balanced each other out, but unfortunately it does not work that way. Because the signals are being driven and measured at different times, and because of variations in the electrode resistances, what actually happens is a chaotic combination of touch and anti-touch. Because energy is being pushed down the X lines and coupled into the Y lines through the water, there tends to be more anti-touch signal than touch signal.

If a finger is placed in the middle of the puddle, two interesting things happen. First, the water couples the finger to the entire surface area it covers. In other words, the area of the finger is artificially expanded to the area of the puddle. Second, the conductive finger is a much lower resistance path to ground than the capacitively coupled Y electrodes. The drive energy that was being cross coupled into the sense lines through retransmission is instead directed into the finger. As a result the anti-touch signals disappear. Also, if any additional fingers make contact with the same puddle, then all those touches merge into one giant touch.

What does all that mean in terms of touch qualification and performance? The anti-touch by itself is not a major problem. The anti-touch signals will not be qualified as a touch (unless the touch controller is using some kind of node-to-node differential algorithm to qualify touches). Depending on the controller and how it is configured, the anti-touches may cause recalibrations or may interfere with the noise detection algorithms, but this can usually be fixed with careful tuning.

The touch signals caused by the puddle could be an issue if the signal levels are as high as the signal level of a finger. This is not usually the case, but it can happen especially with sensors that have a high sensitivity. And it is much more likely to happen with salt water. Again, careful tuning can usually resolve this.

The giant touch signal can cause accuracy issues. Let's say you have a small button on your GUI and there is a puddle of water on the sensor. About 30% of the puddle is covering the displayed button. The other 70% of the puddle is off to the side of the button in an area with no touch elements. The user places their finger on the button and makes contact with the water. The touch controller calculates the centroid of the touch signal. The puddle is stretching out the touch signal so that most of it lies off of the button. As a consequence the reported touch coordinate is not on the button and the software ignores it.

Merged touches can also cause major problems. If two fingers cannot be independently resolved, then no two-finger gestures like pinch, zoom, and rotate will work.

What about a moving finger? This is where things get really interesting. As a finger moves into the puddle, the entire puddle becomes a large touch. What happens next depends on how the water moves with the finger. In most cases, the water will stretch out as the finger moves through it forming a long train connecting back to the original puddle (Fig. 18.2).

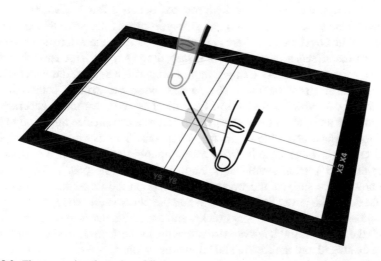

Fig. 18.2 Finger moving through puddle

This creates a moving, smeared rooster tail of touch signals trailing behind the finger as it moves, making it especially difficult for the touch controller to qualify the touch as it forms odd, elongated shapes. Depending on the speed of the moving finger, the way the water spreads out or breaks up, and the scan rate of the sensor, the touch controller may report more than one touch.

In an earlier chapter we talked about hydrophobic coatings. They can be especially useful for reducing false touches from water. But they also create a few problems. If the water is a misting spray, then the hydrophobic coating helps because all of the individual drops from the spray remain separated. Tiny, separated drops do a poor job of coupling one electrode to another, so this greatly reduces the anti-touch and touch signal levels. When a finger touches the sensor, it is only touching a few drops at a time, not one giant puddle, which reduces the footprint of the touch and increases the accuracy of the reported coordinates.

One downside of hydrophobic coatings is what happens when a finger is dragged around the cover lens. Because the surface energy of the hydrophobic coating is very low, any water that comes into contact with the finger tends to cling to the finger. And any water that comes into contact with the water already attached to the finger tends to stick to that growing collection of water. The finger basically becomes a water magnet, dragging a giant puddle of water around with it. This may reduce the rooster tail effect, but it also results in one giant touch that follows the finger wherever it goes. When a finger moves around a cover lens without a hydrophobic coating the water will tend to separate from the finger and then rejoin it, thus reducing the overall amount of water stuck to the touch. Still, the benefits of a hydrophobic coating far outweigh the few problems it creates. The best thing about a hydrophobic coating on a touch surface is that you can brush the water off of the cover lens using your hand and most of it will be removed. Without the coating this just creates smears of water over the entire surface.

If the touch sensor is at an angle, a different set of circumstances occur. First let's consider a cover lens without a hydrophobic coating. Small amounts of water distributed across the surface will tend to stay in place (of course this depends on the angle of the cover lens). If there are only small drops, characteristic of a misting spray, then the effects are limited as the water drops are not large enough to cause cross electrode coupling. As a finger moves around the sensor, the water tends to smear and form larger puddles. The smears cause the same rooster tail effect mentioned above which can result in false touches and poor touch accuracy. But on the positive side, when the water puddles get to a certain size they tend to break free of the surface and slide down the cover lens. This is where a hydrophobic coating can be a huge benefit. With a hydrophobic coating and a decent angle (greater than 45°), even small drops tend to slide off the panel. When a finger moves around the surface, the water that collects on the finger tends to break off and slide down the cover lens as well. When the panel is drenched with water, the water slides off the cover lens very quickly and actually reduces the accumulation of water on the surface. And it is even easier to wipe the surface clean since stuck drops become dislodged and are free to slide down the surface.

Fig. 18.3 Enclosure with
recessed touch sensor

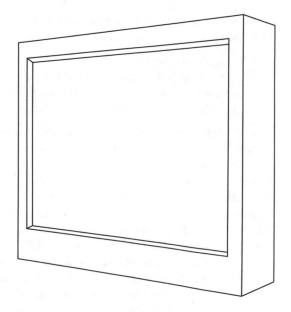

There is of course a down side to this behavior. A large, sliding rivulet of water can look like a touch. In fact, if you were to just look at the raw measurements across the surface of a sensor, it would sometimes be difficult to tell the difference between a wet finger sliding down the cover lens and a large puddle of water sliding down the panel. There are some tuning steps that can help reduce these types of false touches (mainly by looking at the overall shape of the touch signals), but an occasional false touch may still sneak through.

The area most prone to generate false touches is the bottom of the touch sensor. The first reason is that as a puddle of water slides off the sensor the elongated nature of the signals decreases. At some point the touch signals at the edge nodes are about the same size as an actual finger. This can be very difficult to differentiate from an actual touch. One option is to use a debounce-type algorithm, where a touch has to be present in the same spot for several scan cycles before being reported. If the water slides off the cover lens fast enough, then the touch signals will not be stable enough to be pass the debounce criteria.

The second major issue with the bottom edge depends on the bezel design. If the touch sensor is rear mounted and the top surface of the touch sensor module is recessed below the enclosure surface, this creates a ledge around the touch sensor area (Fig. 18.3).

Water running off the surface collects along the bottom edge. As more water rolls off it makes contact with the large puddle of water and forms temporary scattered touches across the bottom of the sensor. If the puddle is large enough, it may cause false touches along the bottom even without additional runoff. And because the water is essentially held in place by the bezel, the touches are relatively stable so a debounce algorithm will not prevent them from being reported.

All of this leads to four guidelines when designing a product that will be used in an environment where water could land on the top surface of the touch sensor:

1. Use a hydrophobic coating on the top surface.
2. Design the product be as close to vertical as possible when being used.
3. Always use a front mount assembly with a flush bezel.
4. Do not place touch elements along the bottom edge of the GUI.

Another prudent option is to have a cleaning mode where all (or most) touches are ignored. There are several ways to implement this. The safest is to use a separate button to enable and disable the cleaning mode. This button could be mechanical or be a PCAP button separate from the touch sensor. Another option is to have a GUI button that activates and deactivates the self cleaning mode. This button should be as high on the GUI as possible (assuming the device is mounted at an angle). Of course the button would have to be active during the cleaning process so the user can touch it when they are done. This raises the possibility that button could be accidentally activated during the cleaning. An alternative is to have a timed cleaning mode. Once the user hits the cleaning button, they have 20 seconds (for example) to wipe off the top surface before the touch sensor is reactivated. Still another option is to automatically disable the touch sensor once a certain number of touches are detected, then automatically re-enable it when the number of touches is zero.

As mentioned in the previous chapter, some touch controllers monitor the raw signal levels and execute a self calibration when the signal levels have too much variance. This can help when the variance is caused by factors like temperature or humidity changes, but it can also cause havoc in applications where water is likely to get on the sensor. If the touch controller sees a large mixture of anti-touch and touch signals caused by water and then recalibrates, the touch sensor will begin behaving erratically. It may generate constant false touches, may not respond to real touches, or both. In applications that require water immunity, it is usually a good idea to disable any auto calibrations that might cause this to happen.

In Chap. 3, we discussed self capacitance and how it can help prevent false touches from water and improve finger tracking through water. The reason self capacitance performs better in the presence of water is because of the self capacitance drive cycle. Typically during a self capacitance scan all of the electrodes are driven simultaneously. Because they are all being driven, retransmission effects are eliminated or greatly reduced (depending on the salinity of the water). If the touch controller uses both mutual and self scans, it can compare the reported touch signals in the mutual scan data to the reported touch signals in the self scan data. If it sees touches in the mutual data but not the self data, the controller ignores the touches. Of course the algorithms are not quite that simple. In fact they can be incredibly complex. Water detection and immunity is an area where a more expensive controller can truly outshine the low cost alternative. If your application requires water immunity, spend the extra money for a touch controller that includes algorithms specifically designed to detect and remove water signatures.

Chapter 19
GUI Design and Gesture Engines

Designing a touch-enabled application is not difficult, but it does require a different way of thinking. Many application designers, when working on their first touch-enabled GUI, tend to design the kind of user interface you would use on a PC application with a mouse. Using a mouse has become such second nature to us that we do not realize what a precise tool it is, or that it allows us to place the location of the intended action (i.e., the tip of the mouse arrow) very carefully, down to an individual pixel, before executing the action. A touch interface works very differently. First, we cannot know for sure where the actual touch report will be located within the area of the tip of our finger. We have a rough idea that it will be somewhere near the center of the finger, but that can change depending on the angle of the finger in relation to the touch sensor. Second, as the finger approaches the surface it blocks our view of the very item we are trying to touch, making it even more difficult to be precise. Both of these issue can be reduced by using a stylus, but even then the preciseness of the touch depends on the size of the stylus and how we hold it. Third, usually the touch down executes an action. This is very different from a mouse where we can move the mouse cursor around the screen without triggering any action.

Fortunately, we all have experience using a touch-enabled device – a cell phone. Look at the way your cell phone's screen is laid out. On the home screen the application buttons are relatively large with a sizeable gap between them. A typical human finger is around 8–10 mm wide. The touch areas on the GUI should be roughly the same size or larger to ensure easy usage. There should also be a 4–6 mm gap between touch areas. Of course these are only guidelines. You can have smaller buttons and smaller gaps, but you risk frustrating and alienating users.

If your display is very small or if your application requires a lot of touch areas on the screen at once, there are some strategies that can help reduce the user's

NOTE: The material in this chapter was first published as an article, "The challenges of multi-touch gesture interfaces", on Embedded.com (https://ubm.io/2IjfbiC).

frustration and improve accuracy. For example, the touch action can execute when the finger is lifted instead of when it makes contact. That allows the user to drag their finger around the GUI and only lift it when they are over the intended target. Further, the touch element itself can change when the touch is on top of it. This helps the user identify which element they are touching. The element might change color, become translucent, shake, grow, or do something else that the user can see. Haptics might also be used to help the user feel the borders between touch elements. There are very few if any haptics solutions that work well enough for this type of feedback today, but improvements are constantly being made in haptic feedback technology.

Many operating systems and GUI libraries include gesture recognition for both single touch and multi-touch gestures. However, if you are using a low-level embedded operating system, then you may have to write your own gesture processor. The following sections provide some guidelines on recognizing and reporting various touch events, starting with single touch only gestures.

The simplest and most common single touch gestures are *tap*, *double tap*, *drag*, and *flick*. A single tap, which is analogous to a click event with a mouse, is defined by the amount of time that a touch is active and the amount of movement during the touch. Typical values might be that the touch down and touch up events must be less than a half second apart, and the finger cannot move by more than 5 pixels. A double tap is a simple extension of the single tap where the second touch must occur within a certain amount of time after the first touch, and the second touch must also follow the same timing and positional requirements as the first touch. Keep in mind that if you are implementing both a single tap and a double tap, then the single tap processing will need to wait a certain amount of time before sending the single tap event to ensure that the user is not executing a double tap.

While a *drag* gesture is fairly simple to implement, it is often not needed at the gesture recognition level. Since the touch controller only reports coordinates when a finger is touching the panel, your application can treat those coordinate reports as a drag. Implementing this at the application level has the added benefit of knowing if the origination of the drag occurred over an element that can be dragged. If not, then the touch reports can be ignored or continually analyzed for other events (e.g., passing over an element may result in some specific behavior).

A *flick* is similar to a drag, but it has a different purpose. A drag event begins when the finger touches the panel and ends when the finger is removed. A flick can continue to generate events after the finger is removed. This can be used to implement the kinds of fast scrolling features common on many cell phones where a list continues to scroll even after the finger is lifted. A flick can be implemented in several ways, with the responsibilities divided between the gesture recognition layer and the application layer. But before we discuss the different ways to implement a flick gesture, let's first focus on how to define a flick.

A flick is generally a fast swipe of the finger across the surface of the touch panel in a single direction. The actual point locations during the flick do not typically matter to the application. The relevant parameters are velocity and direction. To identify a flick, the gesture recognition layer first needs to determine the velocity of the finger movement. This can be as simple as determining the amount of time between the finger down report and the finger up report divided by the distance traveled. However, this also can slow the response time since the velocity is not determined until after the gesture has finished. For greater responsiveness, we can track the velocity as the gesture is occurring. If the velocity is above a certain threshold, then a flick has occurred. The direction of the flick can also be determined from the touch down and touch up positions. However, this simplistic approach can result in false flicks. The user may draw a very quick circle which meets the velocity requirements but most certainly should not be interpreted as a flick. To prevent this type of false report, the gesture recognition engine should determine if all the reported points are in a relatively straight line. If not, then the gesture is not a flick. If the algorithm is reporting the velocity in real time, then it must include some kind of prequalification of the gesture direction before it starts reporting flick events to the application. The gesture recognition engine also needs to decide how to report the direction of the flick. This is driven by the user interface requirements and the amount of decoupling you want between the gesture recognition layer and the application. In the most generic version, a flick might report the direction as an angle. Or it could report it in a more application-friendly way such as left, right, up, or down.

Another typical feature of a flick is inertia. Inertia provides a natural end to a flick operation by, for example, slowing down the scrolling of a list until it comes to a stop. Inertia can be implemented at the application level or the gesture recognition level. At the gesture level, inertia can be implemented as a series of events with decreasing velocity. This could be implemented with specific inertia events, but it is typically easier to reuse the flick event for the same purpose. The initial flick event includes the direction and velocity of the actual flick gesture. A timer in the gesture recognition engine then continues to generate flick events using the original direction and a decreasing velocity. The decreasing velocity is generated by an exponential decay function which is tuned to create whatever deceleration profile works best for your application. You might even want to dynamically control the deceleration profile based on the application state (i.e., "larger" objects might decelerate more quickly than "smaller" ones, or longer lists might scroll for a longer period of time before decelerating).

The C source code for a simple flick algorithm is shown below. This flick algorithm is based on the standard formula for calculating distance traveled using velocity and acceleration:

$$d = vt + \frac{at^2}{2} \tag{19.1}$$

The velocity is the initial velocity of the flick gesture. The acceleration and time are tuned to meet the application's requirements. In this case, we use a constant time and determine the rate of deceleration required to complete the decay in that time. A timer is used to generate each of the flick deceleration events. The timer should fire every 10–200 milliseconds depending on processor utilization and the required smoothness of the flick. On each timer tick we need to move a certain number of pixels, which decreases with each step. The number of pixels to move on each step is calculated by taking the delta between the total distance of the previous step and the total distance of the next step:

$$\text{Delta} = \left(vt_{n+1} + \frac{at_{n+1}^2}{2} \right) - \left(vt_n + \frac{at_n^2}{2} \right) \tag{19.2}$$

The flick processing is divided among three areas of the source code:

```
//////////
// INIT //
//////////
const uint16 FLICK_TIMER_DURATION_IN_MS = 100;
const double MIN_FLICK_TIME_IN_MS = 1000.0;

// calculate the duration of a flick in timer ticks
uint8 NumberOfDecelerationSteps =
    MIN_FLICK_TIME_IN_MS /
    FLICK_TIMER_DURATION_IN_MS;

// round up if needed
if (MIN_FLICK_TIME_IN_MS % FLICK_TIMER_DURATION_IN_MS)
    ++NumberOfDecelerationSteps;

/////////////////////////////
// AT BEGINNING OF FLICK //
/////////////////////////////

// TODO: calculate InitialVelocityInPixelsPerMs

double Deceleration =
    InitialVelocityInPixelsPerMs /
    (2.0 * FLICK_TIMER_DURATION_IN_MS);

uint8 DecelerationStep = 1;

// TODO: start flick timer

/////////////////////////
// EACH TIMER TICK //
/////////////////////////
uint32 CurrentStepTimeSquared =
    pow(DecelerationStep *
        FLICK_TIMER_DURATION_IN_MS, 2);

uint32 PreviousStepTimeSquared =
    pow((DecelerationStep - 1) *
        FLICK_TIMER_DURATION_IN_MS, 2);
```

```
uint32 PixelsToMoveThisStep =
    InitialFlickVelocityInPixelsPerMs *
    FLICK_TIMER_DURATION_IN_MS -
    Deceleration *
    (CurrentStepTimeSquared -
    PreviousStepTimeSquared);

// if this is the last step, disable the flick timer
if (++DecelerationStep > NumberOfDecelerationSteps)
{
    // TODO: disable flick timer
}
```

This code can be modified to use a constant deceleration (instead of constant time) and to use fixed point instead of floating point. When porting and modifying the code (especially when converting to fixed point), be sure that the variable types are large enough to hold the resulting calculations.

The two most common multi-touch gestures are *rotate* and *zoom* (here zoom refers to zooming in or out). In its simplest form a rotation gesture begins once the second finger is down and ends when there is only one finger still on the touch sensor. The center of rotation is typically defined as the midpoint between the fingers. The rotation angle can be reported as a relative angle referenced to the original positions of the two fingers, or it can be reported as an absolute angle referenced to the panel orientation. The gesture engine might even provide both angles and let the application decide which one to use. Some applications might need the actual coordinates of the fingers instead of the angle. This can be useful for sticky tracking of the finger positions. In this mode, the coordinates of each finger act as anchor points for manipulating elements on the user interface. A typical example would be an image-viewing application. When user places their fingers on the image and then moves their fingers around, the application modifies the shape and position of the image as necessary to keep the initially touched pixels directly under the user's fingers.

The rotation angle is based on the position of the two fingers defining the rotation. The coordinates of the two touches can be used in conjunction with the arctangent function to return an angle in radians:

```
AngleInRadians =
    atan((SecondTouch.Y - FirstTouch.Y) /
         (SecondTouch.X - FirstTouch.X));
```

The angle can be easily converted to degrees if needed:

```
AngleInDegrees = AngleInRadians * 180.0 / 3.14159;
```

There are several issues with the atan function as used above. First, if the two X coordinates are the same, then a divide-by-zero error occurs. Second, the atan function cannot determine which quadrant the angle is in and so returns a value between +π/2 (+90°) and –π/2 (−90°). It is better to use the `atan2` function which handles both of these problems and returns a value between π (+180°) and –π (−180°):

```
AngleInRadians =
    atan2(SecondTouch.Y - FirstTouch.Y,
          SecondTouch.X - FirstTouch.X);
```

The use of floating-point libraries can be avoided by creating a table of angles based on the ratio of Y and X. This has the added advantage of avoiding the radians-to-degrees conversion. The Y value will need to be multiplied by some factor so that the resulting Y-to-X ratio can be expressed as an integer. The size of the table and the Y multiplier used depends on the required resolution of the angle. A four entry table is enough to return up, right, down, or left. A table with 360 entries can return the exact angle in whole degrees. Just be sure that the Y multiplier used to calculate the ratio is large enough to give the desired resolution.

The rotation angle is typically reported as an offset from the original angle. The angle of the first two touches becomes the reference angle. As the fingers rotate around the panel, the actual rotation angle is the difference between the reference angle and the current angle.

A *zoom* gesture occurs when the user places two or more fingers on the touch panel and then moves those fingers closer together or further apart. A zoom event might report a positive or negative distance in pixels. As with the drag gesture, it may be easier to implement a zoom gesture directly in the application using the reported touch positions. The application would enter a zoom state whenever two or more fingers are present. Once in the zoom state, the application uses the coordinates of the touches to implement the zoom. This has the added benefit of following the same sticky tracking mentioned above where the pixels under the user's fingers stay with the fingers as they move around the panel. The downside is that the gesture engine must determine if a zoom or rotation is happening before sending the touch information to the application.

In addition to the simple gestures mentioned above, the application may need to recognize combinations of gestures. The most common example is simultaneous rotate and zoom gestures where a user element (like an image or a map) sticks to the user's fingers. If this is the most common kind of gesture manipulation used in the application, a gesture-recognition engine might not be needed since it may be easier for the application to just receive the individual touch coordinates and implement the desired functionality.

Other combinations can be more complicated, like a zoom with flick. In this scenario, the user is trying to zoom out or in very quickly. Just like a flick, a zoom can have an initial velocity and a deceleration curve, either positive for zooming out or negative for zooming in. Again, the gesture engine requirements are entirely driven by the application.

Another added complication is the need to track multiple gestures simultaneously. This is a typical requirement in collaboration applications where more than one user may be interacting with the application. When tracking multiple gestures, the gesture-recognition engine will need to report some kind of gesture identifier for each event. It also needs to keep track of which physical touches are assigned to each gesture. This information must remain constant until all the gestures have ended.

Tracking multiple gestures raises some difficult issues when determining which touch is associated with which gesture. For example, assume that three rotation gestures are occurring simultaneously. Gesture 1 is using physical touches A and B, gesture 2 is using touches C and D, and gesture 3 is using touches E and F. First touch D is lifted, then touch E is lifted, then a new touch is applied. Should this new touch be associated with gesture 2, gesture 3, or does it start a new gesture? Is it possible to assign the new touch to a gesture based on proximity (in other words, it has to be within 20 pixels of the other touch in a gesture)? The answers to these questions can be simple, or can be based on complicated predictive algorithms tailored to the application. Either way, the gesture engine must explicitly deal with these issues to avoid inconsistent and confusing results for the users.

Once the required gesture events have been generated, the next issue is how to efficiently get this information to the application. You can set up your own event system, but the difficulty is in deciding what information to pass up to the application. Good software design states that there should be very little coupling between the application and the gesture-recognition system. However, some types of gesture-recognition engines can be greatly enhanced by allowing information to flow both directions across the application/gesture engine boundary. For example, it may be possible to disambiguate between a drag and a flick if the engine can report the initial location of the gesture to the application and the application can determine if the initial touch occurred on a scroll bar (drag) or in the body of the displayed information (flick).

Of course most of the larger operating systems already include gesture processing engines. Some touch controllers also include limited gesture processing which can be very useful for embedded applications that use a real-time operating system or no operating system. But if neither your OS or touch controller can provide gesture information, then writing some basic gesture processing is not very difficult. The most important thing to remember when designing your gesture engine is that a touch is not the same thing as a mouse click. While this might seem limiting at first, once you begin to truly analyze the user interface from a touch perspective, you will discover an entirely new way to communicate with your users.

Chapter 20
Automated Testing

Now that we have covered how a PCAP touch sensor works, how it is built, how it responds to different environmental factors, and how to integrate it with your operating system, we come to the final chapter: how to test the sensor. Testing the performance of a PCAP touch sensor involves a lot of subtle and not-so-subtle issues including mechanical alignment, coordinate systems, test definitions, dealing with electrical noise, operating system certification, etc. There are several companies that sell automated testing systems for touch sensors. Most of these systems are well designed with easy-to-use software. Many of them include optional add-ons for testing the performance of multiple touches including pinch separation, multi-finger gestures, response time, and more.

So why not just buy one of the off-the-shelf solutions? The main reason is because they tend to be very expensive. Another reason is that most of the available test platforms are geared toward cell phones and tablets. Many either are not capable of supporting larger sizes (15.6 in or bigger) or are a lot more expensive for the larger size option. If you need a wide variety of test capabilities including multiple finger support, or if you are implementing testing for a specific operating system's certification requirements, investing in an off-the-shelf solution is probably worth it. If, however, you just want some basic testing capabilities, especially for large touch sensors, then designing and building your own test system can be much more affordable. Either way, it is helpful to have an understanding of what the test system needs to do and what choices need to be made.

For the purposes of this discussion, we will assume that the automated test system is based on an inexpensive tabletop CNC. The base is a large, flat metal plate that does not move (typically extruded aluminum – see Fig. 20.1).

The "finger" is a metal stylus mounted onto a carriage that moves on a three axis motorized system. The computer can move the carriage to any position on the plate, then raise and lower the stylus.

This is just an example for discussion. On some systems, the plate moves in X and Y and the stylus only moves up and down. Regardless of what moves or how the movement is implemented, the key is that the stylus can be manipulated to make

© Springer International Publishing AG, part of Springer Nature 2019
T. Gray, *Projected Capacitive Touch*, https://doi.org/10.1007/978-3-319-98392-9_20

Fig. 20.1 Tabletop CNC

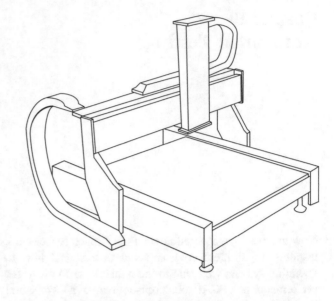

Fig. 20.2 Plastic
base

contact and then break contact with any point on the surface of the Unit Under Test
(UUT).

The first issue to deal with is how the UUT is mounted on the plate. For starters,
there are several good reasons not to place the UUT directly on the metal surface.
First, the metal surface is a great conductor of electrical noise coming from the
motors and the motor driver unit. Second, the metal surface may not be very smooth
and may scratch the UUT. Typically a thick piece of plastic or Lexan® is mounted to
the metal plate and the UUT is placed on top of it (Fig. 20.2).

The next issue is how to secure the UUT to the plastic base. The first option is to
mount a right angle alignment bracket to the plastic base and manually align the
UUT up against it (Fig. 20.3).

The benefits of this arrangement are that only one alignment bracket needs to be
made, it can be permanently mounted to the plate, and it is easy to swap from one
UUT to another. The downsides are that manually placing the UUT up against the
bracket introduces some small initial error, and the edges of the UUT have to be
straight. Another issue is that the height of the alignment brackets needs to be
adjustable if the thickness of the UUTs being tested varies. If the alignment bracket
is too low, then the UUT may not make contact with the bracket, making it difficult
to align the UUT properly (Fig. 20.4).

Fig. 20.3 Alignment bracket

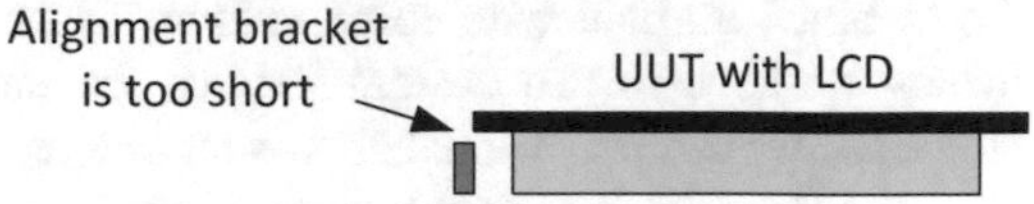

Fig. 20.4 Short alignment bracket

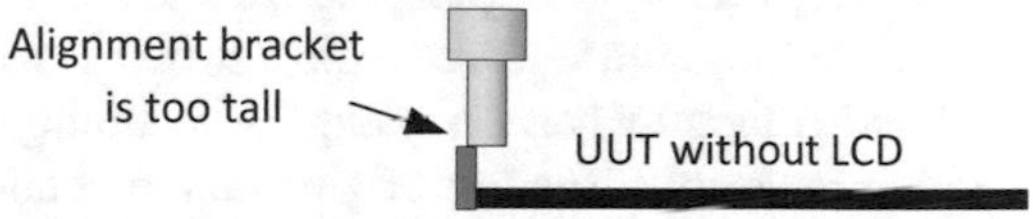

Alignment bracket is too short

UUT with LCD

Fig. 20.5 Tall alignment bracket

Alignment bracket is too tall

UUT without LCD

Fig. 20.6 Sliding bracket

If the alignment bracket is too tall, it may interfere with the stylus being used for testing at the outer edges of the sensor's active area (note that the stylus shown in Fig. 20.5 is much larger than a typical stylus because it is simulating a 10 mm finger).

This problem can be solved by having several alignment brackets of various heights and by mounting the brackets to the side of the plastic base with a sliding adjustment (Fig. 20.6).

The other option for mounting the UUT is to make a custom fixture that mounts directly to the plate. The benefits of using a custom fixture are perfect alignment every time with little effort and the ability to accommodate any size or shape of UUT. The downsides are the cost and lead time for each custom fixture.

However we align the UUT to the table, it is very important that the sensor be aligned to the table's axes. Even a small rotation can introduce a large error especially on larger panels. As an example let's consider a situation where the touch sensor is rotated $0.5°$. If the touch sensor is fairly small, say 4.3 in diagonal, then the lower right corner will be about 0.3 mm away from where it should be. That means that even if the reported touch location in that corner is exactly correct, our test will still report a location error of 0.3 mm at the far corner of the sensor. That is not too bad and is typically within an acceptable accuracy range. But what if we are testing a 15.6 in diagonal touch sensor? In that case a 0.5° rotation equates to an error of 3.4 mm at the far corner. That much error will completely skew the results.

So even a small rotation can equate to a large error at the corner furthest from the test origin. Our first concern is to minimize that rotation error as much as possible. We could add a step to the test software that measures the angle of rotation and halts the test if it is greater than some threshold. Of course there will be some inaccuracy when collecting the reported touches we need to calculate the sensor rotation. Even so we can probably get a good enough set of data if we use test points in the middle of the sensor where the error should be the smallest. We could even be more aggressive and apply a rotation correction to our reported touches, although this is not ideal as we are now adjusting the reported data by what is essentially a fudge factor based on data with some unknown inaccuracy. You might think that this means we have to use custom fixtures for each UUT to guarantee no rotation. However, we can use manual alignment as long as the alignment fixture is perfectly aligned to the table's axes, the mechanical pieces we align to are perfectly made, and we are very careful in aligning the UUT to the fixture. With careful attention to these details, it is possible to achieve a less than 0.1° rotation even with manual alignment.

Another factor when designing the mounting options is whether the test origin is fixed or moveable. The test origin is the first corner of the sensor's active area that the stylus will make contact with. The location of that origin can either be fixed in relation to the base, in which case the position of the UUT needs to be adjusted so that the active area corner is in the correct position, or the starting origin of the test needs to be adjustable, in which case the UUT can be placed just about anywhere. If using alignment brackets, the test origin must be adjustable. In this case determining

Fig. 20.7 UUT test origin with active area

Fig. 20.8 Test origin

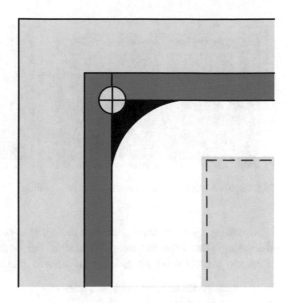

Fig. 20.9 Active area
distances from edges

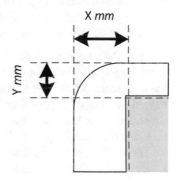

the test origin is actually fairly easy. Let's say that the UUT is mounted as shown in Fig. 20.7 (note: the dotted red line represents the active area of the sensor).

The sensor is aligned against the two mounting brackets. Those brackets do not move; their position is constant. That means that the position of the corner of the active area is also constant for this particular UUT design. In the test software we have a constant location variable that identifies the coordinates for the inside corner of the two brackets. We call that the test origin, identified by the yellow circle and crosshairs in Fig. 20.8.

Now we just have to look at the mechanical drawing for the UUT and measure the distance from the two edges of the cover lens to the active area (Fig. 20.9).

We enter these two values into our test software, and it calculates the coordinates of the corner of the active area.

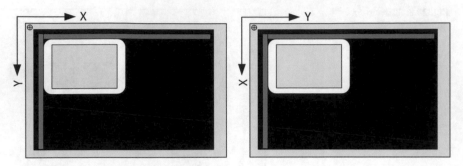

Fig. 20.10 Automated tester coordinate system options

Speaking of coordinates, one of the most confusing issues to deal with when designing your own automated tester is the coordinate system. The four most common coordinate systems that you have to deal with are:

- Automated tester's coordinate system
- Native coordinate system of the UUT
- Configured coordinate system of the UUT
- Monitor coordinates assigned by the operating system

The CNC table (or whatever automated system is being used) has its own coordinate system with its own origin. Typically the origin is the home position of the three motors that control the X, Y, and Z positions. The motors cannot be safely moved past this origin position, so this also represents the minimum starting point of a test. From that home position (shown as a green crosshairs) the X and Y axes could go either direction (Fig. 20.10).

The touch sensor has a native coordinate system and a configured coordinate system. The native coordinate system is defined by the electrode order. Most touch controllers include settable parameters that allow the coordinate system to be swapped and rotated to any desired orientation. This is called the configured coordinate system. The native and the configured coordinate systems could have any of the eight orientations shown in Fig. 20.11.

Finally, there is the monitor's coordinates which are almost always defined from the top left corner of the monitor (Fig. 20.12).

The test software moves the stylus to a location defined in the automated tester's coordinate system. It then lowers the stylus onto the surface of the UUT. The UUT reports a touch to the operating system using the UUT's configured coordinate system. The operating system then transforms the UUT's reported coordinates into its own coordinate system. Ultimately, our goal is to determine whether the reported touch location matches the expected touch location. But we are comparing the stylus location defined in the automated tester's coordinates with the reported touch defined in the operating system's coordinates. How do we compare these two different pairs of coordinates and determine if the touch controller reported the expected location?

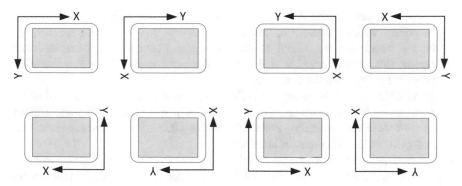

Fig. 20.11 UUT coordinate systems

Fig. 20.12 Operating
system coordinate system

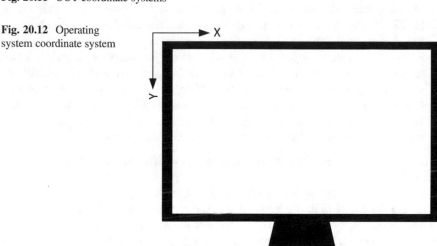

First, let's agree that there are a variety of ways this issue could be solved. For example, the test software could start with an expected coordinate for the touch system, then transform that location into the automated tester's coordinate system for positioning the stylus. Another option is for the test software to transform the stylus coordinates, given in the automated tester's coordinate system, into operating system coordinates. And of course there are several other combinations of transformations that could be used, any of which will allow us to determine if the reported touch coordinate is what we expected. That being said, there are a few preferred options:

1. If possible, it is better to use the UUT's reported coordinates instead of the operating system's coordinates. Typically, the touch controller has a higher resolution than the computer monitor. The higher resolution of the UUT's coordinates is lost when the operating system scales the UUT's coordinates into its coordinate system. In addition to scaling the coordinates, the operating system will most likely have to rotate them as well. It can also be difficult to relate the operating system's coordinates to the touch sensor. For example, if the report

shows one area of the sensor has having much worse accuracy, how do you determine which section of the sensor matches that area? For these reasons, it is best to avoid these additional transformations and the errors they introduce. Of course this is only possible if the test software can directly access the touch reports coming from the controller.

2. If possible, it is better to use the UTT's native coordinate system instead of the configured coordinate system. Again, the issue here is avoiding any unnecessary coordinate transformations, especially if the UUT's configuration includes a reduction in the coordinate resolution.

3. An alternative to #2 is, instead of stripping away the UUT's orientation settings, we can use them to reorient the UUT's coordinate system so that it matches the automated tester's coordinate system. Of course we cannot match actual positions (meaning that coordinate (1123, 456) in one coordinate system is not going to be (1123, 456) in the other coordinate system), but we could at least match up the location of the origin and the directions of the axes.

4. Assuming that we are logging the test data and not just analyzing it once, it is best to save each set of coordinates, the stylus location, and the reported touch location using their original coordinate systems. Of course we have to transform at least one of these to perform our comparison, but we want to keep the original values as well. This gives us a more detailed history of exactly what happened during the test, and it helps us to confirm that the test is working correctly.

After the test is completed, we end up with a table of values like the ones shown in Table 20.1.

In this example the stylus coordinates are in millimeters offset from the origin (meaning the corner of the active area), and the reported touch coordinates are in the native coordinate system of the UUT. For now are also assuming that the origins of the two coordinate systems align and that the axes are defined with the same orientation. To determine how accurate the reported touches are, we need to transform one coordinate system into the other. Again, we could do it either way. But when the time comes to report the results to the user, they would probably prefer that the calculated error in the position be given in units that are meaningful. In this case, the automated system's units, millimeters, will be much more meaningful than the touch coordinates. It also helps that the automated system's units are decimal while the touch coordinates are integer. In general, it is always better to increase the unit resolution (i.e., go from integer to decimal) instead of decreasing it (i.e., go from decimal to integer).

We can transform the reported touch coordinates into the automated tester's coordinate system with a simple scaling of each axis (Eqs. 20.1 and 20.2):

Table 20.1 Stylus and reported coordinates	Stylus coordinates		Reported touch coordinates	
	X	Y	X	Y
	0.0	0.0	8	10
	0.0	2.0	7	85
	0.0	4.0	9	170

Table 20.2 Stylus, reported, and transformed coordinates

Stylus coordinates		Reported touch coordinates		Transformed touch coordinates	
X	Y	X	Y	X	Y
0.0	0.0	8	10	0.29	0.23
0.0	2.0	7	85	0.26	1.97
0.0	4.0	9	170	0.33	3.94

$$\text{Touch } X \text{ in mm} = \text{Touch } X \text{ in touch pixels} * \frac{\text{Length of } X \text{ axis in mm}}{\text{Max } X \text{ coordinate} + 1} \quad (20.1)$$

$$\text{Touch } Y \text{ in mm} = \text{Touch } Y \text{ in touch pixels} * \frac{\text{Length of } Y \text{ axis in mm}}{\text{Max } Y \text{ coordinate} + 1} \quad (20.2)$$

For this example we will assume that the length of the X axis is 150.0 mm, the length of the Y axis is 95.0 mm, and both axes have a max coordinate of 4095. The transformed coordinates are shown in Table 20.2.

For the three reported points shown in the table, our X location was off by an average of 0.29 mm and our Y was off by an average of 0.07 mm. We can apply the same transformations for all of the reported touch coordinates and determine the error for each point, the average error for the entire sensor, the average error in X and Y, the average error close to the edges of the active area, or any other combination of errors we want.

In our example, we assumed that the two coordinates systems were aligned with each other. Most likely they will not be aligned (unless we set up the touch controller's orientation specifically so they match). If they do not match, then we need to apply some additional transformations before we can compare them. Again, it is better to transform the touch coordinates into the automated system's coordinates so that we increase the resolution during the transform instead of decreasing it. The first transformation is to match the X axis of the coordinate system with the X axis of the UUT. This is fairly easy. If they already match we do nothing. If they do not match, we just swap the X and Y coordinates for each touch (X becomes Y and Y becomes X). The second transform is to get each axis to go in the same direction. Again, if they already match then we do nothing. If they do not match, then we just reverse the coordinates. For example, if the X axis is reversed and if the max X coordinate is 4095, then we subtract each X coordinate from 4095 (i.e., New $X = 4095 - \text{Original X}$). After these transformations we convert the touch coordinates from touch pixels to millimeters and compare them to the stylus position.

At this point we understand how to collect data from the automated system, transform it, and use it to calculate the error in the reported position. But what kind of test pattern should we use? How far apart should our test points be? To answer these questions we have to consider several factors:

1. How large is the UUT?
2. How quickly can the stylus be moved to a new location?
3. Are there areas where the error will likely be worse?

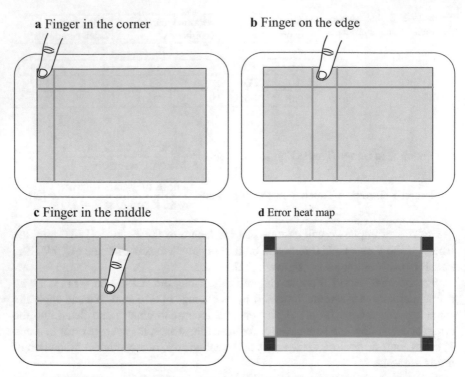

Fig. 20.13 (**a**) Finger in the corner. (**b**) Finger on the edge. (**c**) Finger in the middle. (**d**) Error heat map

If we are testing a small UUT and our system is relatively fast, then we could capture a test point every 1 mm (there is not much point in going small than 1 mm: the error is unlikely to change that quickly within an area and the noise in our reported touch position will most likely be close to ±0.25 mm anyway). If we executed this test for a 7 in sensor which has an active area of 152 mm by 91 mm, then we would capture 13,832 points. If the automated system is capable of moving to a new position within 2 seconds, which is pretty fast, then the total test time would be around 8 hours! And that is only for a 7 in sensor. For a 10.1 in sensor the total test time would be around 16 hours. That is with a relatively fast stylus movement time of 2 seconds. Most CNC tables cannot move that quickly. How can we reduce the test time and still characterize our sensor?

If we look at the typical error pattern of a PCAP touch sensor, we find that the highest error occurs in the corners (Fig. 20.13a). *Note: in the figures the blue lines represent electrodes.* The reason the error is so high here is that there is only one node close to the finger. If the finger is not between two nodes, it is very difficult to calculate its position. The next worst error occurs when the finger is between the outer electrode and the edge of the active area (Fig. 20.13b). In this area the finger has a moderate impact on the two closest nodes. The position between the nodes can be calculated with good accuracy, but the distance to the edge is harder to determine.

When the finger is in the middle of the sensor, meaning when it is anywhere inside the four outer electrodes, it is surrounded by four nodes and the position can be determined with high accuracy (Fig. 20.13c). Figure 20.13d shows the area of worst (red), medium (yellow), and best (green) accuracy in relation to the four outer electrodes (blue).

If we adapt our test resolution to the error map, using a higher resolution of test points in the areas with higher error and a lower resolution in the areas with lower error, we can greatly reduce the number of test points and the length of time the test takes to execute. In theory we could use three different resolutions, but that would overly complicate the test. A good compromise is to use a higher resolution in all of the outer edge areas including the corners and a lower resolution in the inner area. Also, we do not want to stop using our higher resolution right at the outer electrode as the error directly on these electrodes will also be somewhat worse. Another issue is that we do not want to adapt each test to the specific electrode pitch. On most diamond pattern sensors the electrode pitch is between 4 mm and 8 mm. Let's use half of the worst case pitch and say that we want to use our higher resolution anytime we capture a point within 12 mm of the edge. Of course this is somewhat arbitrary. If you are designing your own test system and want to do something similar, feel free to use 10 mm, 15 mm, or whatever suits your sensor designs and test equipment.

So what resolution should we use in these two areas? A good compromise between reducing test time and good test coverage is to use a 2 mm resolution around the edges and a 6 mm resolution in the middle. Given that a typical finger has a diameter of between 8 mm and 10 mm, those values insure that we are always capturing data at a higher resolution than the typical touch. It also has the benefit that 2 divides into 6. It is helpful for the low resolution pitch to be a multiple of the high resolution pitch so that our test points line up in columns and rows. This will help with some of the linearity calculations we will discuss later. Using these numbers on a sensor with a 7 in active area would result in the test pattern shown in Fig. 20.14.

There is a subtle issue hiding in the question of pitch and the starting point of the test. Let's say for example that the long edge of the active area is 137.6 mm and that it runs left to right. We start capturing data points in the upper left corner. We capture the first point, then move 2 mm to the right. We capture another test point and move another 2 mm to the right. We keep this up until we come to the right edge. How are our test points distributed in relationship to that edge? In this case with a 137.6 mm length, our last data point will be captured at a distance of 136.0 mm from the left edge. Our next point would be 138.0 which is past the right edge. In other words, unless our total length is an exact multiple of our pitch, we will not hit the far right edge. So what do we do?

There are several ways to deal with this issue:

1. If the last test point is more than a certain distance from the edge (say 0.25 mm), then move the stylus directly onto the edge and capture a point. The spacing to the last column (and the last row) would not match the rest of the data, but that is not a major issue. It just complicates the software a bit.

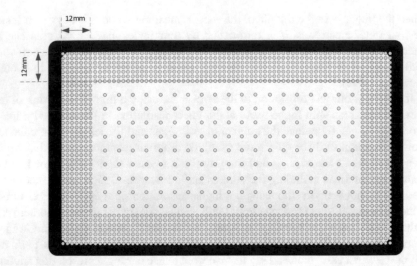

Fig. 20.14 Seven inch test pattern

2. Adjust the actual test point pitch so that we always end up exactly on the edge while still being as close to 2 mm as possible. For this example, our test point pitch would be 1.9942 mm. Of course we would also have to adjust our low resolution pitch to 5.9826 mm so that the rows and columns line up. And we have to be careful that round off accumulation does not cause issues the further we get from the origin.
3. Don't worry about it. Capturing data right on the edge is not going to provide us with much more information than we already have.

Any of these will work. It is just a matter of careful coding of the test and analysis software.

Now that we have our test data, what can we do with it? The first, most obvious answer is to determine the error of the reported locations. Before we get into the details, let's first discuss the terminology. A lot of documents and articles refer to this as accuracy testing. That is true in the sense that we are trying to determine how accurate our touch sensor is. But it is also wrong in the sense that we do not measure accuracy, we measure error. If the reported point is exactly where it should be, then our error is 0. If the reported point is not where it should be, then however we measure error, it is going to increase as the reported point gets further from its expected location. It would be a misnomer to refer this measurement as accuracy since a larger number mean less accurate, not more. For that reason, we are going to use the term "location error" or just error instead of accuracy throughout this discussion.

We could calculate the location error as the linear distance from the expected location to the reported location, as a vector (meaning that we include the angle and direction from the expected location to the reported location), or as an XY offset. These three options are demonstrated in Fig. 20.15.

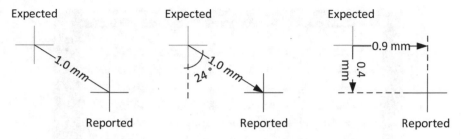

Fig. 20.15 Error reports

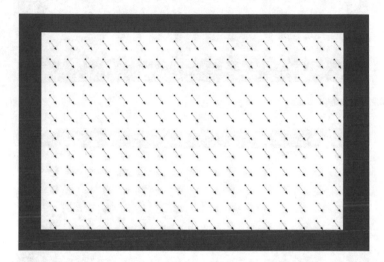

Fig. 20.16 Sensor coordinates are shifted

The vector option can be useful in determining if the error is caused by a misalignment or incorrect tuning of the sensor. For example, if most of the error vectors point down and to the right, then the mechanical starting point for the test may have been misaligned, or the panel may be tuned with some invalid scaling corrections (many controllers include options for scaling the coordinates to correct for poor mechanical alignment). Fig. 20.16 shows an example of what the error vectors might look like when there is a mechanical or scaling shift.

Another useful pattern in the error vectors is if most of the error vectors point toward the center or away from the center of the active area (Fig. 20.17).

This is a really strong indication that some scaling factor is incorrect, either in the touch controller's configuration or in the operating system (if the operating system's coordinates are being used).

If the panel is correctly aligned and there are no scaling errors, then the error vectors should have a somewhat random distribution (Fig. 20.18).

The XY offset can be useful for determining if the error is greater in the vertical direction or the horizontal direction. If there is a large difference between the two,

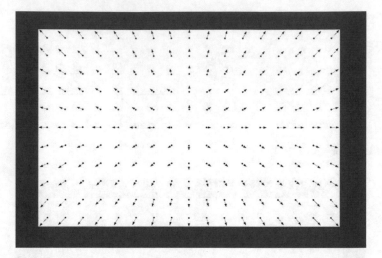

Fig. 20.17 Incorrect scaling

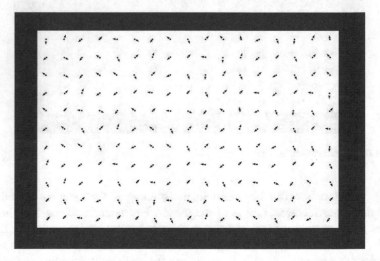

Fig. 20.18 Randomly distributed error vectors

that might also indicate an incorrect mechanical alignment of the tester or an issue with the touch controller's configuration.

When analyzing the error, it is often useful to determine the average error in the edge region (i.e., inside the 12 mm border we are using to separate our high resolution testing from our low resolution testing) versus the average error in the middle. The error around the edges will often be double or more than the error in the middle. If we really want to delve into the performance, we could even analyze the error within a few millimeters of the corners, or the error of the points along the very edges of the active area. All of this can be interesting data especially when comparing one sensor design to another. For example, how much better is the accuracy if the electrode pitch is 4 mm versus 6 mm?

The average error is going to depend on a number of factors including:

- The electrode pitch
- The sensor pattern geometry
- The touch controller
- The ambient electrical noise
- The monitor resolution (if the operating system's coordinates are being used)

Just to give an idea of normal error ranges, an error of ± 0.5 mm in the center is excellent but often unnecessary for most applications. An error of ± 1.0 mm in the center is very good and probably only needed in consumer applications. An error of ± 2.0 mm in the center is usually good enough for all other applications (industrial controls, medical terminals, exercise equipment, etc.). The edge error should be no more than 50% higher than the error in the center. Some controllers include sophisticated edge correction algorithms that can reduce the increased error to only 10% when properly configured. If you are planning on pursuing operating system certification for your device, be sure to read through the certification documentation to understand how it defines accuracy, how accuracy is determined during testing, what the pass criteria are, and if there are different pass criteria near the edges compared to the middle.

In our discussion about test pitch and origin, note that we did not mention the exact electrode pitch or node locations. In fact if we pick a standard test pattern pitch, whether it is 0.5 mm, 1.0 mm, 2.0 mm, or something else, it is unlikely that there will be any correlation between where the test points are captured and the node locations of the electrodes. Some test points may fall directly on top of nodes, but most will not. Does this matter? Generally we will get the best accuracy directly on a node as that provides a single signal peak with a clearly defined location. The worst accuracy (ignoring the corners and edges) is directly in the center of four nodes. The signals of the four surrounding nodes are all affected by the finger, but not equally. First a finger is more oblong than round, so the nodes closer to the longer axis of the finger's ellipse will be more affected, skewing the calculation of the finger's position. Second, the finger is probably at a slight angle. The nodes that lie in the direction of the finger's tilt are going to be more affected and skew the direction calculation. If our test locations are randomly aligned to the nodes, then some test points will be directly on the nodes and naturally have the least error, some will be exactly between four nodes and have the most error, and the rest will have random error levels. If we were interested in understanding this, we could devise a test pattern that is aligned to the node locations. For example, we might pick a center node, then capture test points every 0.25 mm within this node (Fig. 20.19).

There are a number of tests like this that we could devise. For example, we could study how the error changes as we move in 0.25 mm steps out from the corners of the node. Or we could study how the error changes as we move along an entire electrode, or between electrodes, or diagonally, or any other way we want. But ultimately the question we are trying to answer is a simple one: when the user touches the sensor,

Fig. 20.19 Testing a node

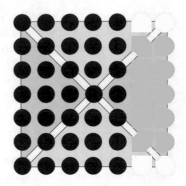

will the reported touch be where they expect it to be? Even if the user's finger is a relatively small 6 mm, they are not going to be able to discern if the reported location is off by less than 1 mm. And if they are trying to touch a button on the GUI that is 12 mm by 12 mm, it is not going to matter if the center of the button is directly over a node or exactly between nodes. Either way the user will not have any problems touching the button. As long as we can measure the error every few millimeters and make sure it is relatively small compared to the application's requirements, that is all we need.

In addition to location error, there is another common attribute that is often specified and measured for PCAP touch sensors: linearity. For a touch sensor, linearity is defined as the deviation (or really the lack of deviation) from a straight line. If we put our stylus down on the panel and draw a perfectly straight line, which is a much easier thing to do with an automated system than with a human finger, how straight is the reported line? As with accuracy vs. location error, the terminology here can be a bit confusing. We want to determine how straight the line is, but what we are really measuring is the deviation from straightness. A higher value means less linear, not more. So like we did for location error, we will use the term "linearity deviation" or just deviation in this section.

We could measure linearity by literally putting the stylus down on the sensor, dragging it from edge to edge in a straight line, and then lifting it again. There are a few difficulties with this kind of test. First, anytime the stylus is making contact with the sensor we are going to get a *lot* of touch reports, typically 100–200 per second. If we just raise and lower the stylus, then it is fairly easy to determine a single coordinate for the touch. We can take the first reported location on touch down, the last reported location on lift, or an average location of multiple reports. But if we drag the stylus across the sensor, it is going to be difficult to correlate the mechanical location of the stylus to the reported location. There will always be some lag between the two. The lag may be small, but any lag adds an unwanted amount of error. The other issue with dragging is determining how much data we want to keep. If we are getting 200 touch reports per second and it takes 10 seconds to drag from the top of the sensor to the bottom, that is 2000 points. If we repeat that every 2 mm on a 150 mm long sensor, we would have to store 150,000 points. We could cut that down by only saving the points every few millimeters. For example, maybe we capture a point every time the mechanical position has changed by exactly 2.000 mm. But then we have the problem of figuring out exactly which touch report to save since we do not know which touch report corresponds to that exact moment when the stylus was 2.000 mm from the last point we saved.

Fortunately, there is an easy way out of this difficulty. We already have a test pattern defined that captures a test point every 2 mm in the edge boundary and every 6 mm in the center. We can use this same data to measure the linearity deviation. We just look at one column or row of data points and treat them as if they formed a single straight line. You might be thinking that if the stylus was moving up and down when those points were captured instead of dragging, then we cannot really tell what the deviation is. In fact we can as long as the alignment between our test pattern and the electrodes is random.

Earlier we talked about the location accuracy changes at different positions relative to the center of a node. A test point that is exactly on top of a node will be very accurate while a point between two nodes will be less accurate. The same thing applies whether the stylus is being dragged or lifted and put back down again. In either case the location error will be greater in between nodes than directly on top of them. As long as our columns and rows are randomly distributed so that some columns (or rows) run right through the center of the nodes, some columns run through the middle of the nodes, some columns are closer to one side of the nodes, and some columns are closer to the other side of the nodes, then we will have a good overall idea of the linearity deviation of the sensor.

This greatly simplifies and speeds up our testing. We only need to capture one set of data, then use that data to determine both the location error and the linearity deviation. The one caveat to that is that we can only measure the linearity deviation in rows or columns. Measuring linearity deviation when moving diagonally can be useful as a diagonal movement constantly changes the distance between the touch location and the node centers. But unless we setup our test pattern *very* carefully, we will not have a set of test points that we can follow in a straight line from one corner to another. But again, as long as our rows and columns are random compared to the electrode pitch, then we can still get a useful measurement of the linearity deviation using the row and column data.

How do we actually measure the linearity deviation? For each row and column, we determine the expected touch location for that row or column. For example, if our columns are parallel to the sensor's X axis, then every point in that column has the same expected Y coordinate. For each point, we calculate the deviation in the Y coordinate only. We then average all of those deviations to determine the average deviation for that column. We repeat the same process for each row using the expected X coordinate (Fig. 20.20).

Once we have collected our test data, we can analyze the location error and linearity deviation in a number of different ways using the methods outlined above. We can compare the performance along the edges to the performance in the middle, compare the performance of the X axis to the Y axis, compare the performance of a panel with a 3 mm cover lens to one with a 1 mm cover lens, compare the performance of a 3 mm electrode pitch to a 6 mm electrode pitch, and so on.

Fig. 20.20 Linearity error

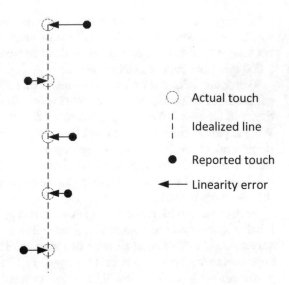

There are a number of possible ways to present the accuracy error and linearity deviation data. One of the best visualization techniques for presenting a large amount of data is to show it graphically. In this case we are especially interested in how the error changes in different areas of the sensor. The center of the sensor should have the lowest error. How do the outer edges and corners compare to the center? The vector graphs shown earlier (Figs. 20.16, 20.17 and 20.18) are excellent ways to show the accuracy error. The lengths of the vectors tell us how the accuracy error changes in different areas of the sensor. The directions of the error vectors can tell us if there are setup issues with alignment, scaling, or rotation of the sensor. They can also show certain patterns in the errors. For example, due to the edge issues mentioned before, the error vectors around the edges tend to point inward from the edges. If the touch controller includes edge correction features, then comparing the edge error vectors with and without edge correction enabled tells us if our edge correction is setup correctly. For linearity we can draw straight lines and show error vectors coming off those lines. We can quickly see if the linearity is worse vertically or horizontally, or closer to the edges vs. further from the edges. And of course we can also create tables of accuracy and linearity figures, compare the average errors for different parts of the sensor, calculate statistical data like standard deviations, etc. This brings us to the final point about testing: do not get buried in it! Our ultimate goal is to ensure that when a user can touch a specific element on the screen. As we have mentioned several times, fingers are relatively large and clumsy pointing devices. Styli are only slightly better. Our touch sensor does not have to be accurate to the tenth of a millimeter for the user to have a positive experience.

There are several reasons that projected capacitive touch sensors have gone from essentially 0% market share in 2006 to dominating the market today. PCAP sensors have better accuracy and linearity than most other touch technologies (especially resistive). More importantly, their accuracy and linearity do not change over time.

But accuracy and linearity are not the main reasons that PCAP has taken over the consumer market and is quickly expanding into the industrial, medical, exercise, appliances, and other markets. PCAP sensors offer true multi-touch functionality. Multi-touch is possible with other touch technologies, although PCAP generally does it better with more touches and higher accuracy. But even multi-touch does not explain the exponential rise of PCAP over the last decade.

The real reason that PCAP sensors are being used on almost every touch enabled device today is because they work with a "projected" capacitive field. Using an electric field to sense touch through a piece of glass or plastic allows the device to have a perfectly flat top service and be completely sealed. When Apple created the first iPhone, they created an iconic industrial design that took advantage of projected capacitive touch technology. Within just a few years of the launch of the original iPhone, companies began insisting that their products have the "iPhone look," meaning flat, sleek, and with no (or limited) mechanical buttons. As a result, users now expect a touch interface on every display they come into contact with. That huge shift in human behavior never would have happened without the technology behind the projected capacitive touch sensor.

Index

Printed in the United States
By Bookmasters